Analysis

Foundations of Differential and Integral Calculus

Lucien Sina

4.7.2026

Contents

Chapter 1

Preface

Analysis is one of the fundamental areas of mathematics and is also of central importance for many other mathematically oriented fields of study. In particular, for students of mathematics, computer science, physics, and engineering, it provides an indispensable foundation for further study.

This book is aimed primarily at students in their first two semesters, especially at those who encounter analysis systematically for the first time in their first semester. Its focus is a careful introduction to the central concepts and methods of analysis. These include, among others, mathematical induction, the properties of the real numbers, sequences and series, continuity, differentiation, integration, as well as more advanced topics such as uniform convergence, power series, and Taylor series.

The presentation places particular emphasis on mathematical precision. Concepts are introduced exactly, connections are developed carefully, and statements are proved consistently. At the same time, care has been taken to present the material not only formally, but also in an intuitive and comprehensible way. Many examples, remarks, and exercises are intended to help readers not merely read the material, but truly understand it.

The book is structured so that it can be used both alongside a lecture course and for independent study. Detailed derivations and numerous exercises with solutions are intended to make analysis more accessible and to enable readers to work through and consolidate the material systematically.

My aim was to write a textbook that takes the mathematical content seriously without making it seem unnecessarily intimidating. For many students, analysis is at first a demanding subject, but with growing understanding it gains clarity and structure. If this book helps to make that access easier and perhaps even conveys some enjoyment of mathematics, then it has fulfilled its purpose.

Lucien Sina

Chapter 2

Mathematical Induction and Combinatorics

Analysis deals with continuous change, limiting processes, series, sequences, functions, and the question of how a general law for infinitely many cases can arise from finitely many steps. For precisely this reason, it is helpful to begin the study of analysis with a proof principle that formalizes exactly this kind of step-by-step reasoning: mathematical induction.

With its help, statements about all natural numbers can be proved by first establishing the initial case and then showing that the validity of any given case implies the validity of the next one. In this way, a single secure starting point leads to a universally valid conclusion. This principle appears repeatedly in analysis, for instance in summation formulas, inequalities, recursions, estimates for sequences, and many other statements that involve a natural number as a parameter.

Mathematical induction is therefore not merely a technical tool, but a central way of thinking in mathematics. Anyone who masters it reliably has an important foundation for understanding and developing analysis.

2.1 Proof by Mathematical Induction

The principle of mathematical induction is based on a fundamental property of the natural numbers: Every natural number is either the beginning of the sequence or is obtained from a smaller natural number by adding 1. This structure is exactly what makes it possible to derive statements for all natural numbers from an initial case and an inheritance step.

Definition 1 (Principle of mathematical induction). *Let $M \subseteq \mathbb{N}_0$. If*

1. $0 \in M$, *and*

2. *for all $n \in \mathbb{N}_0$, the implication holds: if $n \in M$, then $n + 1 \in M$,*

then already $M = \mathbb{N}_0$.

This principle can also be formulated as a general statement about properties:

Definition 2 (Induction principle in statement form). *Let $P(n)$ be a statement about $n \in \mathbb{N}_0$. If*

1. *$P(0)$ is true, and*
2. *for every $n \in \mathbb{N}_0$, the validity of $P(n)$ implies the validity of $P(n+1)$,*

then $P(n)$ is true for all $n \in \mathbb{N}_0$.

A proof by mathematical induction therefore always consists of two essential parts:

1. **Base case:** One shows that the statement holds for the first relevant value of the natural number.
2. **Induction step:** One assumes that the statement holds for an arbitrary but fixed n, and then shows that it also holds for $n + 1$.

The induction step is the decisive transition. It shows not only that a single case is true, but that the truth of the statement propagates along the natural numbers. Once the starting point has been secured, every subsequent case is automatically covered.

This is called *complete* induction because individual numbers are not checked in isolation; instead, the entire infinite system of natural numbers is captured by a general logical structure. In analysis, this method is particularly important, since many statements must be proved for all $n \in \mathbb{N}$ or $n \in \mathbb{N}_0$.

Remark 3. *Depending on the convention, the natural numbers begin with* 0 *or with* 1*. The form used here,* $\mathbb{N}_0$*, contains* 0*. If a later chapter works with* $\mathbb{N} = \{1, 2, 3, \dots\}$*, only the base case changes, not the underlying principle.*

The principle of complete induction reveals its strength especially when it is applied to a concrete statement. A classical example is the sum of the first n natural numbers. This example makes it particularly clear how a proof by induction is structured and why the transition from n to $n + 1$ forms the real core of the method.

Example 4 (Small Gaussian summation formula)**.** *One of the first and at the same time best-known applications of mathematical induction is the summation formula*

$$1 + 2 + \cdots + n = \frac{n(n+1)}{2} \qquad (n \in \mathbb{N}).$$

It describes the sum of the first n *natural numbers and is therefore often called the* small Gaussian summation formula.

At first glance, the left-hand side seems to require a new calculation for each new value of n*. The formula shows, however, that this sum can be transformed into a simple closed form. Statements of this kind are typical in analysis: from a sequence of individual values, a general law emerges.*

Proof. We prove the formula by mathematical induction on n.

For clarity, we denote the sum of the first n natural numbers by

$$S_n := 1 + 2 + \cdots + n.$$

Thus we have to show that

$$S_n = \frac{n(n+1)}{2} \quad \text{for all } n \in \mathbb{N}.$$

1. Base case. For $n = 1$, we have

$$S_1 = 1.$$

On the other hand, the right-hand side gives

$$\frac{1 \cdot 2}{2} = 1.$$

Hence the statement is true for the initial value $n = 1$.

2. Induction step. Now assume that the formula already holds for a fixed but arbitrary $n \in \mathbb{N}$. That is,

$$S_n = \frac{n(n+1)}{2}.$$

This assumption is called the *induction hypothesis.* It does not mean that we want to prove the statement only for this single value of n; rather, we assume it as already known in order to derive the next case from it.

We now consider the sum up to $n + 1$:

$$S_{n+1} = 1 + 2 + \cdots + n + (n+1).$$

Since $1+2+\cdots+n = S_n$, the induction hypothesis gives

$$S_{n+1} = S_n + (n+1) = \frac{n(n+1)}{2} + (n+1).$$

Now we factor out $n+1$:

$$S_{n+1} = (n+1)\left(\frac{n}{2}+1\right) = (n+1)\left(\frac{n+2}{2}\right).$$

Thus

$$S_{n+1} = \frac{(n+1)(n+2)}{2}.$$

This is exactly the desired formula for the case $n+1$.

Since the base case has been proved and validity for n always implies validity for $n+1$, the statement is proved for all $n \in \mathbb{N}$. □

This example illustrates the basic pattern of complete induction particularly clearly: first, a starting point is established, and then it is shown that each case implies the next one. In this way, a statement can be proved not merely for some numbers, but for all natural numbers.

2.2 Summation Notation

When many terms are added in analysis, it would be impractical to write out every sum in full. For this reason, a special notation for sums is used, the so-called *summation notation*. It makes it possible to write long additions in a clear and general form.

Definition 5 (Summation notation). *Let* $a_1, a_2, \ldots, a_n$ *be numbers and let* $n \in \mathbb{N}$. *Then*

$$\sum_{i=1}^{n} a_i := a_1 + a_2 + \cdots + a_n.$$

The number i *is called the* running index *or* summation index. *It serves only to organize the individual summands and is not regarded as an independent quantity within the sum.*

With the help of summation notation, many statements can be formulated very compactly. Sums whose terms depend on a natural number occur especially often. Precisely in such situations, complete induction is a particularly suitable method of proof.

In the following, we prove a classical example that also plays an important role in working with sums.

Example 6 (Sum of the first odd numbers). *For all* $n \in \mathbb{N}$, *one has*

$$\sum_{i=1}^{n} (2i - 1) = n^2.$$

In other words, the sum of the first n *odd numbers is always equal to the square of* n.

Proof. We prove the statement by complete induction on n.

Base case: For $n = 1$, we have

$$\sum_{i=1}^{1}(2i-1) = 2 \cdot 1 - 1 = 1$$

and at the same time

$$1^2 = 1.$$

Thus the assertion is true for $n = 1$.

Induction hypothesis: Suppose that, for a fixed but arbitrary $n \in \mathbb{N}$, the statement already holds, that is,

$$\sum_{i=1}^{n}(2i-1) = n^2.$$

Induction step: We now show that the statement then also holds for $n+1$. To this end, consider

$$\sum_{i=1}^{n+1}(2i-1).$$

We separate off the last summand:

$$\sum_{i=1}^{n+1}(2i-1) = \sum_{i=1}^{n}(2i-1) + (2(n+1)-1).$$

Using the induction hypothesis, it follows that

$$\sum_{i=1}^{n+1}(2i-1) = n^2 + 2n + 1.$$

Since

$$n^2 + 2n + 1 = (n+1)^2$$

we obtain

$$\sum_{i=1}^{n+1}(2i-1) = (n+1)^2.$$

This proves the induction step.

Consequently,

$$\sum_{i=1}^{n}(2i-1) = n^2$$

for all $n \in \mathbb{N}$. □

Complete induction and many of the examples considered so far are based on the natural numbers. We therefore now introduce the most important number sets, which will be used constantly throughout analysis.

Definition 7 (Natural numbers). *The set of natural numbers is*

$$\mathbb{N} := \{1, 2, 3, \dots\}.$$

If one also includes the number 0, *one obtains the set*

$$\mathbb{N}_0 := \{0, 1, 2, 3, \dots\}.$$

The natural numbers are used in particular for counting and ordering. Because of their step-by-step structure, they form the basis of complete induction.

In order to form differences of natural numbers without restriction, the natural numbers are extended to the integers.

Definition 8 (Integers). *The set of integers is*

$$\mathbb{Z} := \{\dots, -3, -2, -1, 0, 1, 2, 3, \dots\}.$$

Thus it contains the positive and the negative whole numbers as well as 0.

Remark 9. *Complete induction is a proof method and not a definition of the natural numbers. However, it is closely connected with the natural numbers, since they are arranged in a successive sequence-like structure. It is precisely this structure that makes the transition from n to $n+1$ possible in the induction step.*

Further number sets will be introduced later, in particular the rational numbers $\mathbb{Q}$, the real numbers $\mathbb{R}$, and the complex numbers $\mathbb{C}$. For the foundations of complete induction, however, the natural numbers are sufficient at first.

2.3 Product Notation and Factorials

Product notation and factorials are not only useful abbreviations for repeated multiplication, but also central concepts in combinatorics. This area of mathematics studies how many possibilities certain arrangements and selection processes allow. The factorial arises naturally in this context, because it describes the number of all possible orderings of a finite set. Just as summation notation represents long additions compactly, the *product notation* is used to write longer multiplications clearly and concisely. It plays an important role both in analysis and in combinatorics.

Definition 10 (Product notation). *Let $a_1, a_2, \ldots, a_n$ be numbers and let $n \in \mathbb{N}$. Then*

$$\prod_{i=1}^{n} a_i := a_1 \cdot a_2 \cdot \ldots \cdot a_n.$$

The number i is again called the running index *or* multiplication index. *It serves only to arrange the factors in order.*

Analogously to sums, one also considers the empty product. It is useful to define it as 1, since 1 is the neutral element of multiplication.

Remark 11 (Empty product). *For $n = 0$, one sets*

$$\prod_{i=1}^{0} a_i := 1.$$

This convention is not arbitrary; rather, it makes many formulas valid in a uniform way and without exceptional cases.

A particularly important application of product notation is the *factorial.*

Definition 12 (Factorial). *For $n \in \mathbb{N}_0$, the factorial is defined by*

$$n! := \prod_{k=1}^{n} k.$$

In particular, this gives

$$1! = 1, \qquad 2! = 1 \cdot 2, \qquad 3! = 1 \cdot 2 \cdot 3, \qquad \dots$$

and, because of the convention for the empty product, also

$$0! = 1.$$

In combinatorics, the factorial counts the number of all possible arrangements of n distinct elements.

Theorem 13 (Arrangements of an n-element set)**.** *Let M be a set with exactly n elements. Then there are exactly $n!$ different arrangements of the elements of M.*

Proof. We justify the statement step by step.

For the first element of an arrangement, there are n possibilities, since each of the n elements may stand in first position. Once the first element has been chosen, $n-1$ elements remain, so there are $n-1$ possibilities for the second position. For the third position, $n-2$ possibilities remain, and so on.

Continuing in this way up to the last position, one obtains altogether

$$n \cdot (n-1) \cdot (n-2) \cdots 2 \cdot 1 = n!.$$

Thus an n-element set has exactly $n!$ different arrangements. □

Example 14. *A set with three elements has exactly*

$$3! = 3 \cdot 2 \cdot 1 = 6$$

different arrangements. If the elements are, for instance, a, b, c, *then these arrangements are*

$$abc,\ acb,\ bac,\ bca,\ cab,\ cba.$$

Remark 15. *The factorial grows very quickly. Already for* $n = 5$, *one has* $5! = 120$, *and for larger numbers the values soon become very large. This is precisely why the factorial is important in analysis, for example in series, estimates, and later also in Taylor expansions.*

2.4 The Binomial Coefficient

In combinatorics, the same question arises again and again: How many ways are there to choose some elements from a finite set without taking their order into account? The binomial coefficient was introduced precisely for counting problems of this kind. It is one of the most important numerical concepts in elementary combinatorics and later plays a central role in analysis, for example in the binomial theorem and in series expansions.

Definition 16 (Binomial coefficient)**.** *For* $n \in \mathbb{N}_0$ *and* $k \in \mathbb{N}_0$ *with* $k \leq n$, *the* binomial coefficient *is defined by*

$$\binom{n}{k} := \frac{n!}{k!(n-k)!}.$$

For $k > n$, *one additionally sets*

$$\binom{n}{k} := 0.$$

The binomial coefficient $\binom{n}{k}$ is read as "n choose k." It counts the number of ways to choose exactly k elements from an n-element set when the order of selection does not matter.

Theorem 17 (Pascal's recursion formula). *For all $n \in \mathbb{N}$ and all $k \in \{1, \ldots, n-1\}$, one has*

$$\binom{n}{k} = \binom{n-1}{k-1} + \binom{n-1}{k}.$$

Proof. We give a combinatorial proof.

Consider a set M with n elements and fix a particular element $x \in M$. We want to count all k-element subsets of M. By definition, this number is precisely $\binom{n}{k}$.

Now we divide all k-element subsets of M into two disjoint cases:

1. The subset contains the element x.

2. The subset does not contain the element x.

In the first case, in addition to x, one must choose $k-1$ further elements from the remaining $n-1$ elements. This is possible in $\binom{n-1}{k-1}$ ways.

In the second case, x is not chosen, so all k elements must come from the remaining $n-1$ elements. This is possible in $\binom{n-1}{k}$ ways.

Since the two cases are mutually exclusive and together cover all possibilities, the total number of k-

element subsets is

$$\binom{n-1}{k-1} + \binom{n-1}{k}.$$

On the other hand, by definition this total number is $\binom{n}{k}$. Hence

$$\binom{n}{k} = \binom{n-1}{k-1} + \binom{n-1}{k}.$$

This proves the assertion. □

Remark 18. *Pascal's formula is a recurrence relation: it expresses one binomial coefficient in terms of two neighboring binomial coefficients. It is therefore especially useful for constructing tables of binomial coefficients step by step.*

The combinatorial meaning of the binomial coefficient is shown by the following fundamental theorem. It explains why the expression

$$\binom{n}{k}$$

describes precisely the number of k-element subsets of an n-element set.

Theorem 19 (Number of k-element subsets). *Let M be a set with exactly n elements, and let $k \in \{0, 1, \ldots, n\}$. Then there are exactly*

$$\binom{n}{k}$$

distinct k-element subsets of M.

Proof. We count the k-element subsets of M.

First, we do not consider the subsets themselves, but ordered selections of k distinct elements from M. For the first element there are n possibilities, for the second still $n-1$, then $n-2$, and so on, until for the k-th element $n-k+1$ possibilities remain. Altogether, there are therefore

$$n \cdot (n-1) \cdot (n-2) \cdots (n-k+1) = \frac{n!}{(n-k)!}$$

such ordered selections.

However, each k-element subset is counted several times in this way, namely exactly as often as its k elements can be arranged. There are $k!$ different arrangements.

Therefore, one obtains the number of k-element subsets by dividing the number of ordered selections by $k!$:

$$\frac{1}{k!} \cdot \frac{n!}{(n-k)!} = \frac{n!}{k!(n-k)!}.$$

By the definition of the binomial coefficient, this is precisely

$$\binom{n}{k}.$$

Thus an n-element set has exactly $\binom{n}{k}$ distinct k-element subsets. □

2.5 The Binomial Theorem

When powers of a binomial $(a + b)$ are expanded, the resulting expressions become longer as the exponent increases. What is remarkable, however, is that these expressions are formed according to a very regular pattern. This pattern is described precisely by the binomial theorem. It is one of the most important formulas in algebra and also plays a central role in analysis, for example in series expansions and estimates.

If the corresponding coefficients are arranged row by row, one obtains *Pascal's triangle.* Each row contains the binomial coefficients, and each interior entry is obtained as the sum of the two entries above it. Pascal's recurrence formula explains exactly this rule.

Remark 20 (Pascal's triangle). *The binomial coefficients can be arranged in the following way:*

$$\binom{0}{0}$$

$$\binom{1}{0} \quad \binom{1}{1}$$

$$\binom{2}{0} \quad \binom{2}{1} \quad \binom{2}{2}$$

$$\binom{3}{0} \quad \binom{3}{1} \quad \binom{3}{2} \quad \binom{3}{3}$$

and so on.

At the boundary one always has

$$\binom{n}{0} = \binom{n}{n} = 1,$$

and for the interior entries the recurrence formula

$$\binom{n}{k} = \binom{n-1}{k-1} + \binom{n-1}{k} \qquad (1 \leq k \leq n-1).$$

These two properties completely describe Pascal's triangle.

After the binomial coefficients have been introduced as counting numbers, their second central meaning now appears: they are precisely the coefficients in the expansion of a power of $(a+b)$. The following theorem describes this connection.

Theorem 21 (Binomial theorem). *Let $a, b \in \mathbb{R}$ and $n \in \mathbb{N}$. Then*

$$(a+b)^n = \sum_{k=0}^{n} \binom{n}{k} a^{n-k} b^k.$$

Proof. We prove the statement by mathematical induction on n.

Base case: For $n = 1$, we have

$$(a+b)^1 = a + b.$$

On the other hand, the right-hand side gives

$$\sum_{k=0}^{1} \binom{1}{k} a^{1-k} b^k = \binom{1}{0} a^1 b^0 + \binom{1}{1} a^0 b^1 = a + b.$$

Thus the statement is true for $n = 1$.

Induction hypothesis: Assume that, for a fixed but arbitrary $n \in \mathbb{N}$,

$$(a+b)^n = \sum_{k=0}^{n} \binom{n}{k} a^{n-k} b^k.$$

Induction step: We now prove the claim for $n + 1$. To this end, we write

$$(a+b)^{n+1} = (a+b)^n (a+b).$$

Using the induction hypothesis, it follows that

$$(a+b)^{n+1} = \left(\sum_{k=0}^{n} \binom{n}{k} a^{n-k} b^k \right) (a+b).$$

Expanding gives

$$(a+b)^{n+1} = \sum_{k=0}^{n} \binom{n}{k} a^{n+1-k} b^k + \sum_{k=0}^{n} \binom{n}{k} a^{n-k} b^{k+1}.$$

In the second sum, we perform the change of index $k \mapsto k - 1$:

$$\sum_{k=0}^{n} \binom{n}{k} a^{n-k} b^{k+1} = \sum_{k=1}^{n+1} \binom{n}{k-1} a^{n+1-k} b^k.$$

Hence we obtain

$$(a+b)^{n+1} = a^{n+1} + \sum_{k=1}^{n} \left(\binom{n}{k} + \binom{n}{k-1} \right) a^{n+1-k} b^k + b^{n+1}.$$

By Pascal's recurrence formula, for $1 \leq k \leq n$,

$$\binom{n}{k} + \binom{n}{k-1} = \binom{n+1}{k}.$$

Moreover,

$$a^{n+1} = \binom{n+1}{0} a^{n+1} b^0 \quad \text{and} \quad b^{n+1} = \binom{n+1}{n+1} a^0 b^{n+1}.$$

Therefore,

$$(a+b)^{n+1} = \sum_{k=0}^{n+1} \binom{n+1}{k} a^{n+1-k} b^k.$$

This proves the induction step.

Thus the formula holds for all $n \in \mathbb{N}$. □

Remark 22. *The binomial theorem also explains why the binomial coefficients in Pascal's triangle are precisely the correct coefficients in the expansion of* $(a+b)^n$*. Thus, the structure of the triangle directly reflects the algebra of expanding powers.*

2.5.1 Consequences of the Binomial Theorem

By choosing the parameters appropriately, the binomial theorem immediately yields important identities for binomial coefficients. Two particularly well-known formulas arise by setting a and b equal to 1, respectively to 1 and -1. At the same time, this shows how closely algebra and combinatorics are connected.

The binomial theorem provides not only a formula for expanding $(a+b)^n$, but also several important identities for binomial coefficients. These become especially transparent when special values are substituted for a and b. Together with the combinatorial interpretation of binomial coefficients, this leads to two fundamental summation formulas.

Theorem 23. *For all $n \in \mathbb{N}_0$, one has*

$$\sum_{k=0}^{n} \binom{n}{k} = 2^n.$$

Moreover, for all $n \in \mathbb{N}$, one has

$$\sum_{k=0}^{n} (-1)^k \binom{n}{k} = 0.$$

Proof. We apply the binomial theorem.

Substituting $a = 1$ and $b = 1$ into

$$(a+b)^n = \sum_{k=0}^{n} \binom{n}{k} a^{n-k} b^k$$

gives

$$(1+1)^n = \sum_{k=0}^{n} \binom{n}{k} 1^{n-k} 1^k.$$

Since $1^{n-k} = 1$ and $1^k = 1$, it follows that

$$2^n = \sum_{k=0}^{n} \binom{n}{k}.$$

This proves the first assertion.

Now substituting $a = 1$ and $b = -1$ gives

$$(1 + (-1))^n = \sum_{k=0}^{n} \binom{n}{k} 1^{n-k}(-1)^k.$$

Hence

$$0^n = \sum_{k=0}^{n} (-1)^k \binom{n}{k}.$$

For $n \in \mathbb{N}$, we have $0^n = 0$, and therefore

$$\sum_{k=0}^{n} (-1)^k \binom{n}{k} = 0.$$

This proves the second assertion as well. □

Remark 24 (Combinatorial interpretation of the first identity). *The identity*

$$\sum_{k=0}^{n} \binom{n}{k} = 2^n$$

can also be understood directly from a combinatorial point of view. By the theorem on the number of k-element subsets, $\binom{n}{k}$ gives the number of subsets with exactly k elements of a set with n elements. Summing over all possible values of k therefore counts all subsets of an n-element set.

Since each of the n elements is either contained in a subset or not, there are exactly two choices for each element. Hence an n-element set has precisely

$$2^n$$

subsets in total. Thus

$$\sum_{k=0}^{n} \binom{n}{k} = 2^n$$

is also immediately plausible from a combinatorial perspective.

2.6 The Geometric Series

A particularly important special case of sums arises when the summands are generated by repeatedly multiplying by the same number. Such series are called *geometric series*. They occur in many areas of mathematics, for example in estimates, in the study of limits, in interest calculations, and later also in power series.

Theorem 25 (Formula for the geometric series). *Let $a \in \mathbb{R}$ with $a \neq 1$, and let $n \in \mathbb{N}_0$. Then*

$$\sum_{k=0}^{n} a^k = \frac{1 - a^{n+1}}{1 - a}.$$

Proof. We prove the statement by complete induction on n.

Base case: For $n = 0$, we have

$$\sum_{k=0}^{0} a^k = a^0 = 1.$$

On the other hand,

$$\frac{1-a^{0+1}}{1-a} = \frac{1-a}{1-a} = 1.$$

Thus the statement is true for $n = 0$.

Induction hypothesis: Assume that, for some fixed but arbitrary $n \in \mathbb{N}_0$, the formula already holds, that is,

$$\sum_{k=0}^{n} a^k = \frac{1-a^{n+1}}{1-a}.$$

Induction step: We now prove the formula for $n+1$. We have

$$\sum_{k=0}^{n+1} a^k = \sum_{k=0}^{n} a^k + a^{n+1}.$$

Using the induction hypothesis, it follows that

$$\sum_{k=0}^{n+1} a^k = \frac{1-a^{n+1}}{1-a} + a^{n+1}.$$

We put the terms over a common denominator:

$$\frac{1-a^{n+1}}{1-a} + a^{n+1} = \frac{1-a^{n+1}+a^{n+1}(1-a)}{1-a}.$$

After expanding, we obtain

$$\frac{1-a^{n+1}+a^{n+1}-a^{n+2}}{1-a} = \frac{1-a^{n+2}}{1-a}.$$

Hence

$$\sum_{k=0}^{n+1} a^k = \frac{1 - a^{n+2}}{1 - a}.$$

Thus the statement has also been proved for $n + 1$.

Therefore, the formula holds for all $n \in \mathbb{N}_0$. □

Remark 26. *For $a = 1$, the formula is not defined in this form. In this case, however, we immediately have*

$$\sum_{k=0}^{n} 1^k = \sum_{k=0}^{n} 1 = n + 1.$$

The general formula illustrates particularly well how a closed form can arise from a simple repeated structure.

Later, the geometric series will be important, among other things, for limits, series, and estimates. It is therefore one of the fundamental tools of analysis.

The geometric series forms a fitting conclusion to this chapter. It shows once again how effective mathematical induction is as a proof principle and how closely it is connected with counting, summation, and the algebraic structure of the natural numbers. Thus the essential foundations have been laid on which the further study of analysis builds.

2.7 Exercises and Solutions

The following exercises revisit the central contents of this chapter. Each subsection contains exactly one exercise with its solution given immediately afterwards. In this way, the concepts and proof ideas can be reviewed step by step.

1.1 Proof by Mathematical Induction

Exercise 27. *Prove by mathematical induction:*

$$1 + 3 + 5 + \cdots + (2n - 1) = n^2 \qquad (n \in \mathbb{N}).$$

Solution 28. *We prove the statement by mathematical induction on n.*

For $n = 1$, we have

$$1 = 1^2.$$

Thus the assertion is true for the initial value.

Now assume that the statement holds for a fixed $n \in \mathbb{N}$, that is,

$$1 + 3 + 5 + \cdots + (2n - 1) = n^2.$$

Then, for the next case, it follows that

$$1 + 3 + 5 + \cdots + (2n - 1) + (2n + 1) = n^2 + (2n + 1).$$

Since

$$n^2 + 2n + 1 = (n + 1)^2$$

holds, we obtain

$$1 + 3 + 5 + \cdots + (2n + 1) = (n + 1)^2.$$

Thus the assertion has been proved for $n + 1$. Hence the formula holds for all $n \in \mathbb{N}$.

1.2 Summation Notation

Exercise 29. *Write out the sum*

$$\sum_{k=1}^{4}(2k-1)$$

explicitly and compute its value.

Solution 30. *The summation sign means that the terms for $k = 1, 2, 3, 4$ are added:*

$$\sum_{k=1}^{4}(2k-1) = 1+3+5+7.$$

Adding these numbers gives

$$1+3+5+7 = 16.$$

Therefore,

$$\sum_{k=1}^{4}(2k-1) = 16.$$

This also agrees with the general formula

$$\sum_{k=1}^{n}(2k-1) = n^2,$$

since for $n = 4$ we have $4^2 = 16$.

1.3 Product Notation and Factorials

Exercise 31. *Compute*

$$\prod_{k=1}^{5} k$$

and explain its connection with the factorial.

Solution 32. *The product notation means that the factors for* $k = 1, 2, 3, 4, 5$ *are multiplied:*

$$\prod_{k=1}^{5} k = 1 \cdot 2 \cdot 3 \cdot 4 \cdot 5 = 120.$$

By definition, precisely this expression is the factorial of 5:

$$5! = 1 \cdot 2 \cdot 3 \cdot 4 \cdot 5 = 120.$$

Therefore,

$$\prod_{k=1}^{5} k = 5!.$$

1.4 The Binomial Coefficient

Exercise 33. *Compute*

$$\binom{5}{2}.$$

Then briefly explain the combinatorial meaning of the result.

Solution 34. *Using the definition of the binomial coefficient, we obtain*

$$\binom{5}{2} = \frac{5!}{2!\,3!}.$$

Now

$$5! = 120, \qquad 2! = 2, \qquad 3! = 6,$$

so

$$\binom{5}{2} = \frac{120}{2 \cdot 6} = 10.$$

This means that from a set with 5 elements, one can choose exactly 10 different 2-element subsets when the order does not matter.

1.5 The Binomial Theorem

Exercise 35. *Expand* $(x+y)^3$ *using the binomial theorem.*

Solution 36. *For* $n = 3$*, the binomial theorem states that*

$$(x+y)^3 = \sum_{k=0}^{3} \binom{3}{k} x^{3-k} y^k.$$

The binomial coefficients are

$$\binom{3}{0} = 1, \qquad \binom{3}{1} = 3, \qquad \binom{3}{2} = 3, \qquad \binom{3}{3} = 1.$$

Therefore, we obtain

$$(x+y)^3 = x^3 + 3x^2y + 3xy^2 + y^3.$$

This is the complete expansion of $(x+y)^3$*.*

1.5.1 Consequences of the Binomial Theorem

Exercise 37. *Show that*

$$\sum_{k=0}^{n} \binom{n}{k} = 2^n.$$

Solution 38. *We substitute* $a = 1$ *and* $b = 1$ *into the binomial theorem. Then*

$$(1+1)^n = \sum_{k=0}^{n} \binom{n}{k} 1^{n-k} 1^k.$$

Since every power of 1 *is again* 1*, the right-hand side simplifies to*

$$\sum_{k=0}^{n} \binom{n}{k}.$$

On the left-hand side, we have

$$(1+1)^n = 2^n.$$

Hence,

$$\sum_{k=0}^{n} \binom{n}{k} = 2^n.$$

This formula also has a natural combinatorial interpretation: it counts all subsets of a set with n *elements.*

1.5.2 The Geometric Series

Exercise 39. *Show that for* $a \neq 1$ *and* $n \in \mathbb{N}_0$*:*

$$\sum_{k=0}^{n} a^k = \frac{1 - a^{n+1}}{1 - a}.$$

Solution 40. *We prove the formula by complete induction on n.*

For $n = 0$, we have

$$\sum_{k=0}^{0} a^k = a^0 = 1.$$

The right-hand side gives

$$\frac{1 - a^{0+1}}{1 - a} = \frac{1 - a}{1 - a} = 1.$$

Thus the statement is true for $n = 0$.

Now assume that the formula holds for some fixed $n \in \mathbb{N}_0$, that is,

$$\sum_{k=0}^{n} a^k = \frac{1 - a^{n+1}}{1 - a}.$$

Then for $n + 1$ we obtain:

$$\sum_{k=0}^{n+1} a^k = \sum_{k=0}^{n} a^k + a^{n+1}.$$

Using the induction hypothesis, it follows that

$$\sum_{k=0}^{n+1} a^k = \frac{1 - a^{n+1}}{1 - a} + a^{n+1}.$$

Bringing the terms to a common denominator gives

$$\frac{1 - a^{n+1} + a^{n+1}(1 - a)}{1 - a} = \frac{1 - a^{n+2}}{1 - a}.$$

Hence the formula also holds for $n+1$. *Therefore it is proved for all* $n \in \mathbb{N}_0$.

This formula is important because it transforms a long sum into a closed form. Later, it will play a central role in limits, series, and estimates.

Chapter 3

The Field Axioms of the Real Numbers

In the preceding sections, we have seen how mathematical statements can be formulated precisely and proved with the help of logical methods. We now turn to one of the central number systems of analysis, the real numbers $\mathbb{R}$.

The real numbers do not merely form a set on which one can add and multiply; they also possess a number of fundamental properties that describe these operations in a characteristic way. These properties are called the *field axioms*.

We first regard $\mathbb{R}$ as a set equipped with two operations

$$+\colon \mathbb{R} \times \mathbb{R} \to \mathbb{R}, \qquad (x, y) \mapsto x + y,$$

and

$$\cdot\colon \mathbb{R} \times \mathbb{R} \to \mathbb{R}, \qquad (x, y) \mapsto x \cdot y.$$

The first section is devoted to the axioms of addition.

3.1 The Axioms of Addition

Addition on the real numbers satisfies the following properties:

1. **Associative law:** For all $x, y, z \in \mathbb{R}$, one has
$$(x + y) + z = x + (y + z).$$

2. **Commutative law:** For all $x, y \in \mathbb{R}$, one has
$$x + y = y + x.$$

3. **Existence of an identity element:** There exists an element $0 \in \mathbb{R}$ with the property that, for all $x \in \mathbb{R}$,
$$x + 0 = x.$$

4. **Existence of additive inverses:** For every $x \in \mathbb{R}$, there exists an element $-x \in \mathbb{R}$ such that
$$x + (-x) = 0.$$

The identity element of addition is called *zero*; the inverse element belonging to x is called the *opposite number* or the *additive inverse* of x.

3.1.1 Consequences of the Addition Axioms

From the addition axioms one can already derive several fundamental properties that will be used repeatedly throughout analysis. Many later calculations and proofs are based on these simple consequences.

1.

Theorem 41. *The zero element is uniquely determined.*

Proof. Let 0 and $0'$ be two neutral elements with respect to addition. Then

$$0 = 0 + 0'$$

by the neutrality of $0'$. On the other hand,

$$0 + 0' = 0'$$

by the neutrality of 0. Hence $0 = 0'$. Thus the neutral element is uniquely determined. □

2.

Theorem 42. *For every $x \in \mathbb{R}$, the additive inverse is uniquely determined.*

Proof. Let y and z be two additive inverses of x. Then

$$x + y = 0 \qquad \text{and} \qquad x + z = 0.$$

We add z to both sides of the first equation. This gives

$$(x + y) + z = 0 + z.$$

By associativity and commutativity of addition, it follows that

$$(x + z) + y = z.$$

Since $x + z = 0$, we obtain $0 + y = z$, and therefore $y = z$. □

3.

Theorem 43. *We have* $-0 = 0$.

Proof. By definition, -0 is an additive inverse of 0. But since

$$0 + 0 = 0$$

also holds, 0 itself is an additive inverse of 0. By the uniqueness of the additive inverse, it follows that

$$-0 = 0.$$

□

4.

Theorem 44. *The equation* $a + x = b$ *has exactly one solution, namely* $x = b - a$.

Proof. First we prove existence. If we set

$$x = b - a = b + (-a),$$

then, using associativity and commutativity of addition, we obtain

$$a + x = a + (b + (-a)) = (a + (-a)) + b = 0 + b = b.$$

Thus $x = b - a$ is a solution.

To prove uniqueness, let x be any solution of $a + x = b$. Then

$$a + x = b.$$

Adding $-a$ to both sides gives

$$(-a) + (a + x) = (-a) + b.$$

By associativity it follows that

$$((-a) + a) + x = (-a) + b,$$

hence

$$0 + x = (-a) + b.$$

Therefore

$$x = (-a) + b = b + (-a) = b - a.$$

Hence the solution is uniquely determined. □

5.

Theorem 45. *We have* $-(-x) = x$.

Proof. By definition, $-x$ is the additive inverse of x, so

$$x + (-x) = 0.$$

By commutativity of addition, we also have

$$(-x) + x = 0.$$

Thus x is an additive inverse of $-x$. By uniqueness of the additive inverse, it follows that

$$-(-x) = x.$$

□

6.

Theorem 46. *We have* $-(x + y) = -x - y$.

Proof. We show that $(-x) + (-y)$ is an additive inverse of $x + y$. Indeed,

$$\begin{aligned}(x + y) + ((-x) + (-y)) &= (x + (-x)) + (y + (-y)) \\ &= 0 + 0 \\ &= 0.\end{aligned}$$

Thus $(-x) + (-y)$ is an additive inverse of $x + y$. By the uniqueness of the additive inverse, it follows that

$$-(x + y) = (-x) + (-y).$$

This is usually written in the abbreviated form

$$-(x+y) = -x - y.$$

□

3.2 Multiplication Axioms and the Distributive Law

Besides addition, $\mathbb{R}$ also has a second operation, multiplication,

$$\cdot : \mathbb{R} \times \mathbb{R} \to \mathbb{R}, \qquad (x, y) \mapsto x \cdot y.$$

This operation also satisfies fundamental axioms that largely determine arithmetic with real numbers. Together with the distributive law, they describe the algebraic structure of the real numbers as a field.

Multiplication on $\mathbb{R}$ satisfies the following properties:

1. **Associative law:** For all $x, y, z \in \mathbb{R}$,

$$(x \cdot y) \cdot z = x \cdot (y \cdot z).$$

2. **Commutative law:** For all $x, y \in \mathbb{R}$,

$$x \cdot y = y \cdot x.$$

3. **Existence of an identity element:** There exists an element $1 \in \mathbb{R}$ such that for all $x \in \mathbb{R}$,

$$x \cdot 1 = x.$$

4. **Existence of multiplicative inverses:** For every $x \in \mathbb{R}$ with $x \neq 0$, there exists an element $x^{-1} \in \mathbb{R}$ such that
$$x \cdot x^{-1} = 1.$$

5. **Distributive law:** For all $x, y, z \in \mathbb{R}$,
$$x \cdot (y + z) = x \cdot y + x \cdot z.$$
By the commutativity of multiplication, we also have
$$(x + y) \cdot z = x \cdot z + y \cdot z.$$

The identity element of multiplication is called *one*; the inverse element associated with a nonzero element x is called the *reciprocal* or *multiplicative inverse* of x.

3.2.1 Consequences

From the multiplication axioms and the distributive law, a number of fundamental statements follow which will be used constantly in what follows. As with the addition axioms, we shall henceforth use these consequences without further justification.

1.

Theorem 47. *The element one is uniquely determined.*

Proof. Let 1 and $1'$ be two identity elements with respect to multiplication. Then

$$1 = 1 \cdot 1'$$

because $1'$ is an identity element. On the other hand,

$$1 \cdot 1' = 1'$$

because 1 is an identity element. Hence $1 = 1'$. Thus the multiplicative identity element is uniquely determined. □

2.

Theorem 48. *For every $x \in \mathbb{R}$ with $x \neq 0$, the multiplicative inverse is uniquely determined.*

Proof. Let y and z be two multiplicative inverses of x. Then

$$xy = 1 \qquad \text{and} \qquad xz = 1.$$

It follows that

$$y = y \cdot 1 = y(xz) = (yx)z = (xy)z = 1 \cdot z = z.$$

Therefore the multiplicative inverse of x is uniquely determined. □

3.

Theorem 49. *For all $a, b \in \mathbb{R}$ with $a \neq 0$, the equation*

$$ax = b$$

has exactly one solution, namely

$$x = a^{-1}b = \frac{b}{a}.$$

Proof. We first prove existence. Set

$$x = a^{-1}b,$$

then, by associativity, we obtain

$$ax = a(a^{-1}b) = (aa^{-1})b = 1 \cdot b = b.$$

Hence $x = a^{-1}b$ is a solution.

To prove uniqueness, let x be an arbitrary solution of $ax = b$. Then

$$ax = b.$$

Multiplication by a^{-1} gives

$$a^{-1}(ax) = a^{-1}b.$$

By associativity, it follows that

$$(a^{-1}a)x = a^{-1}b,$$

hence

$$1 \cdot x = a^{-1}b.$$

Therefore

$$x = a^{-1}b.$$

Thus the solution is uniquely determined. □

4.

Theorem 50. *For all $x, y, z \in \mathbb{R}$, one has*

$$(x + y)z = xz + yz.$$

Proof. From the distributive law we first obtain

$$z(x + y) = zx + zy.$$

By commutativity of multiplication, $z(x + y) = (x + y)z$, and also $zx = xz$ and $zy = yz$. Therefore

$$(x + y)z = xz + yz.$$

□

5.

Theorem 51. *For all $x \in \mathbb{R}$, one has*

$$x \cdot 0 = 0.$$

Proof. Since $0 + 0 = 0$, the distributive law gives

$$x \cdot 0 = x \cdot (0 + 0) = x \cdot 0 + x \cdot 0.$$

Adding the additive inverse of $x \cdot 0$ to both sides yields

$$0 = x \cdot 0.$$

Hence $x \cdot 0 = 0$. □

6.

Theorem 52. *For all* $x, y \in \mathbb{R}$, *one has*

$$xy = 0 \iff x = 0 \text{ or } y = 0.$$

Proof. If $x = 0$ or $y = 0$, then $xy = 0$ follows immediately from the preceding statement.

Conversely, suppose that $xy = 0$. If $x \neq 0$, we multiply the equation $xy = 0$ by x^{-1} from the left. This gives

$$x^{-1}(xy) = x^{-1}0.$$

By associativity and the preceding statement, it follows that

$$(x^{-1}x)y = 0,$$

hence

$$1 \cdot y = 0,$$

and therefore $y = 0$.

Thus, altogether,

$$xy = 0 \implies x = 0 \text{ or } y = 0.$$

Together with the first implication, this proves the claim. □

7.

Theorem 53. *For all* $x \in \mathbb{R}$, *one has*

$$-x = (-1)x.$$

Proof. First, we compute

$$x + (-1)x = 1x + (-1)x = (1 + (-1))x = 0 \cdot x = 0.$$

Thus $(-1)x$ is an additive inverse of x. By the uniqueness of the additive inverse, it follows that

$$-x = (-1)x.$$

□

8.

Theorem 54. *For all $x, y \in \mathbb{R}$, one has*

$$(-x)(-y) = xy.$$

Proof. By the preceding statement, we obtain

$$(-x)(-y) = ((-1)x)((-1)y).$$

By associativity and commutativity of multiplication, it follows that

$$((-1)x)((-1)y) = (-1)(-1)xy.$$

It therefore remains to show that $(-1)(-1) = 1$. This follows from

$$\begin{aligned} 0 &= (-1) \cdot 0 \\ &= (-1)(1 + (-1)) \\ &= (-1) \cdot 1 + (-1) \cdot (-1) \\ &= -1 + (-1)(-1) \end{aligned}$$

Hence

$$(-1)(-1) = 1.$$

Consequently,

$$(-x)(-y) = xy.$$

□

9.

Theorem 55. *For $x \in \mathbb{R}$ with $x \neq 0$, one has*

$$(x^{-1})^{-1} = x.$$

Proof. By definition, x^{-1} is a multiplicative inverse of x, hence

$$xx^{-1} = 1.$$

By the commutativity of multiplication, we also have

$$x^{-1}x = 1.$$

Thus x is a multiplicative inverse of x^{-1}. By the uniqueness of the multiplicative inverse, it follows that

$$(x^{-1})^{-1} = x.$$

□

10.

Theorem 56. *For $x, y \in \mathbb{R}$ with $x \neq 0$ and $y \neq 0$, one has*

$$(xy)^{-1} = x^{-1}y^{-1}.$$

Proof. It suffices to show that $x^{-1}y^{-1}$ is a multiplicative inverse of xy. Indeed,

$$(xy)(x^{-1}y^{-1}) = (xx^{-1})(yy^{-1}) = 1 \cdot 1 = 1.$$

Hence $x^{-1}y^{-1}$ is a multiplicative inverse of xy. By the uniqueness of the multiplicative inverse, it follows that

$$(xy)^{-1} = x^{-1}y^{-1}.$$

□

3.3 General Associative Law

The associative law makes it possible to write sums and products consisting of more than two terms without ambiguity. At first, an expression such as

$$x_1 + x_2 + \cdots + x_n \qquad \text{or} \qquad x_1 \cdot x_2 \cdots x_n$$

appears to depend on a choice of parentheses, but the following result shows that every complete parenthesization has the same value. Hence the notations introduced above,

$$\sum_{k=1}^{n} x_k \qquad \text{and} \qquad \prod_{k=1}^{n} x_k$$

are well-defined.

Theorem 57. *Let* $x_1, \ldots, x_n \in \mathbb{R}$ *with* $n \geq 2$. *Then every complete parenthesization of*

$$x_1 + x_2 + \cdots + x_n$$

has the same value. The analogous statement holds for the product

$$x_1 \cdot x_2 \cdots x_n.$$

Proof. We first prove the statement for addition by induction on n.

For $n = 2$, there is nothing to prove. For $n = 3$, the assertion is precisely the associative law:

$$(x_1 + x_2) + x_3 = x_1 + (x_2 + x_3).$$

Now suppose that the statement has already been proved for some fixed $n \geq 3$. Consider a complete parenthesization of

$$x_1 + x_2 + \cdots + x_{n+1}.$$

Every such parenthesization decomposes the expression into two completely parenthesized partial sums, which are then added. By the induction hypothesis, the values of these partial sums do not depend on their internal parenthesization. For the remaining addition, the associative law yields the same number independently of the outer parenthesization. Thus the statement is proved for all $n \geq 2$.

For products, the proof is entirely analogous. Instead of the additive associative law, one uses the multiplicative associative law

$$(x_1 \cdot x_2) \cdot x_3 = x_1 \cdot (x_2 \cdot x_3).$$

Consequently, every complete parenthesization of a product also has the same value. □

3.4 General Commutative Law

The commutative law states that, for two summands or two factors, changing their order does not affect the value of the expression. For sums and products with several terms, it follows that the individual terms may also be rearranged in any order. More precisely, the value of a sum or a product does not depend on the order of its summands or factors.

Theorem 58. *Let* $x_1, \ldots, x_n \in \mathbb{R}$ *with* $n \geq 2$, *and let* π *be a permutation of* $\{1, \ldots, n\}$. *Then*

$$x_1 + x_2 + \cdots + x_n = x_{\pi(1)} + x_{\pi(2)} + \cdots + x_{\pi(n)}$$

and

$$x_1 \cdot x_2 \cdots x_n = x_{\pi(1)} \cdot x_{\pi(2)} \cdots x_{\pi(n)}.$$

Proof. We first prove the statement for sums. Every permutation can be written as a sequence of interchanges of adjacent elements. It therefore suffices to show that interchanging two adjacent summands does not change the value of the sum.

Consider a sum of the form

$$x_1 + \cdots + x_i + x_{i+1} + \cdots + x_n.$$

By the general associative law, we may write it as

$$(x_1 + \cdots + x_{i-1}) + (x_i + x_{i+1}) + (x_{i+2} + \cdots + x_n)$$

Using the commutative law of addition, we have

$$x_i + x_{i+1} = x_{i+1} + x_i.$$

Hence

$$(x_1 + \cdots + x_{i-1}) + (x_{i+1} + x_i) + (x_{i+2} + \cdots + x_n)$$

has the same value. Thus, interchanging adjacent summands does not change the value of the sum. Since every permutation can be composed of such interchanges, the assertion follows for sums.

For products, the proof is completely analogous. Here too, by the general associative law, adjacent factors may be grouped together, and by the commutative law of multiplication we have

$$x_i \cdot x_{i+1} = x_{i+1} \cdot x_i.$$

Thus, interchanging adjacent factors also does not change the value of the product. This proves the statement. □

3.4.1 Double Sums and Double Products

The general commutative law not only allows the order of individual summands or factors to be interchanged. It also shows that, for finite sums and products, the order of summation or multiplication does not affect the result. This leads to the rules for double sums and double products.

Theorem 59. *Let* $a_{ij} \in \mathbb{R}$ *for* $1 \leq i \leq m$ *and* $1 \leq j \leq n$. *Then*

$$\sum_{i=1}^{m} \sum_{j=1}^{n} a_{ij} = \sum_{j=1}^{n} \sum_{i=1}^{m} a_{ij}$$

and

$$\prod_{i=1}^{m} \prod_{j=1}^{n} a_{ij} = \prod_{j=1}^{n} \prod_{i=1}^{m} a_{ij}.$$

Proof. We first consider the sum. The left-hand side means that, for each fixed i, one first sums over all j and then sums over the indices i. Altogether, exactly the mn numbers a_{ij} are added.

By the general associative and commutative laws, however, the value of a finite sum is independent of the order and the bracketing in which the summands are combined. Therefore, the same mn summands may just as well be grouped in the reversed order. This gives precisely the right-hand side.

For products, the argument is completely analogous. Here too, the value of a finite product is independent of the order and bracketing of the factors. Hence, likewise,

$$\prod_{i=1}^{m}\prod_{j=1}^{n} a_{ij} = \prod_{j=1}^{n}\prod_{i=1}^{m} a_{ij}.$$

□

3.5 The General Distributive Law

The distributive law states that a product can be distributed over a sum. For two sums, repeated expansion means that each term of the first sum is multiplied by each term of the second sum. This is precisely what the general distributive law describes.

Theorem 60. *Let $m, n \in \mathbb{N}$ and let $x_1, \ldots, x_m, y_1, \ldots, y_n \in \mathbb{R}$. Then*

$$\left(\sum_{i=1}^{m} x_i\right)\left(\sum_{j=1}^{n} y_j\right) = \sum_{i=1}^{m}\sum_{j=1}^{n} x_i y_j.$$

By the general commutative law, this is equivalent to

$$\sum_{j=1}^{n}\sum_{i=1}^{m} x_i y_j.$$

Proof. We prove the statement by induction on n.

For $n = 1$, the claim becomes

$$\left(\sum_{i=1}^{m} x_i\right) y_1 = \sum_{i=1}^{m} x_i y_1,$$

which is exactly the already known generalization of the distributive law.

Now suppose that the statement has already been proved for some fixed $n \geq 1$. Then

$$\left(\sum_{i=1}^{m} x_i\right)\left(\sum_{j=1}^{n+1} y_j\right) = \left(\sum_{i=1}^{m} x_i\right)\left(\sum_{j=1}^{n} y_j + y_{n+1}\right).$$

By the distributive law, it follows that

$$\left(\sum_{i=1}^{m} x_i\right)\left(\sum_{j=1}^{n} y_j + y_{n+1}\right) = \left(\sum_{i=1}^{m} x_i\right)\left(\sum_{j=1}^{n} y_j\right) + \left(\sum_{i=1}^{m} x_i\right) y_{n+1}.$$

By the induction hypothesis,

$$\left(\sum_{i=1}^{m} x_i\right)\left(\sum_{j=1}^{n} y_j\right) = \sum_{i=1}^{m}\sum_{j=1}^{n} x_i y_j.$$

Moreover, by the generalization of the distributive law once again,

$$\left(\sum_{i=1}^{m} x_i\right) y_{n+1} = \sum_{i=1}^{m} x_i y_{n+1}.$$

Combining these identities gives

$$\left(\sum_{i=1}^{m} x_i\right)\left(\sum_{j=1}^{n+1} y_j\right) = \sum_{i=1}^{m}\sum_{j=1}^{n+1} x_i y_j.$$

Thus the claim is proved. □

3.6 Powers

In addition to addition and multiplication, exponentiation also plays a central role. Powers arise from repeatedly multiplying a number by itself. To ensure that this notation is unambiguously defined for arbitrary natural exponents, we first introduce it recursively.

Definition 61. *Let $x \in \mathbb{R}$. For $n \in \mathbb{N}_0$, we define the power x^n recursively by*

$$x^0 := 1$$

and, for $n \geq 0$, by

$$x^{n+1} := x^n \cdot x.$$

In particular,

$$0^0 := 1$$

is therefore fixed by definition.

If $x \neq 0$, then for $n \in \mathbb{N}$ we define the negative powers by

$$x^{-n} := (x^{-1})^n.$$

Thus, for every $x \in \mathbb{R}$ and every $n \in \mathbb{N}_0$, the power x^n is uniquely defined. If $x \neq 0$, then x^n is also defined for all integer exponents $n \in \mathbb{Z}$.

3.6.1 Rules of Calculation for Powers

The most important laws of exponents can be derived directly from the recursive definition of powers. These rules will be used frequently in what follows.

Theorem 62. *Let $x \in \mathbb{R}$ and let $m, n \in \mathbb{N}_0$. Then*

$$x^n \cdot x^m = x^{n+m}.$$

Proof. We prove the statement by induction on m.

For $m = 0$, we have

$$x^n \cdot x^0 = x^n \cdot 1 = x^n = x^{n+0}.$$

Now assume that the assertion has already been proved for a fixed $m \in \mathbb{N}_0$. Then, by the recursive definition of powers,

$$x^n \cdot x^{m+1} = x^n \cdot (x^m \cdot x).$$

Using the associative law of multiplication, we obtain

$$x^n \cdot (x^m \cdot x) = (x^n \cdot x^m) \cdot x.$$

By the induction hypothesis, it follows that

$$(x^n \cdot x^m) \cdot x = x^{n+m} \cdot x = x^{n+m+1}.$$

Thus

$$x^n \cdot x^{m+1} = x^{n+(m+1)}.$$

This proves the assertion. □

Theorem 63. *Let $x \in \mathbb{R}$ and let $m, n \in \mathbb{N}_0$. Then*

$$(x^n)^m = x^{nm}.$$

Proof. We prove the statement by induction on m.

For $m = 0$,

$$(x^n)^0 = 1 = x^0 = x^{n \cdot 0}.$$

Now assume that the assertion has already been proved for a fixed $m \in \mathbb{N}_0$. Then

$$(x^n)^{m+1} = (x^n)^m \cdot x^n.$$

By the induction hypothesis, it follows that

$$(x^n)^m \cdot x^n = x^{nm} \cdot x^n.$$

Using the preceding law of exponents, we obtain

$$x^{nm} \cdot x^n = x^{nm+n} = x^{n(m+1)}.$$

Hence also

$$(x^n)^{m+1} = x^{n(m+1)}.$$

This proves the assertion. □

Theorem 64. *Let $x, y \in \mathbb{R}$ and let $n \in \mathbb{N}_0$. Then*

$$x^n \cdot y^n = (xy)^n.$$

Proof. We prove the statement by induction on n.

For $n = 0$, we have

$$x^0 \cdot y^0 = 1 \cdot 1 = 1 = (xy)^0.$$

Now assume that the assertion has already been proved for a fixed $n \in \mathbb{N}_0$. Then we obtain

$$x^{n+1} \cdot y^{n+1} = (x^n \cdot x)(y^n \cdot y).$$

Using the associativity and commutativity of multiplication, it follows that

$$(x^n \cdot x)(y^n \cdot y) = (x^n \cdot y^n) \cdot (xy).$$

By the induction hypothesis, $x^n \cdot y^n = (xy)^n$. Hence

$$(x^n \cdot y^n) \cdot (xy) = (xy)^n \cdot (xy) = (xy)^{n+1}.$$

Therefore,

$$x^{n+1} \cdot y^{n+1} = (xy)^{n+1}.$$

This proves the assertion. □

3.7 Generalization and Examples

So far, we have formulated the field axioms using the real numbers as an example. In algebra, however, it is often useful to consider not only $\mathbb{R}$, but an arbitrary set K equipped with two operations

$$+\colon K \times K \to K, \qquad \cdot\colon K \times K \to K$$

provided that these operations satisfy the same properties as they do for the real numbers. Such a set is called a *field.*

Thus the real numbers are only one example among several. In the following, we mention some important fields.

- **The rational numbers $\mathbb{Q}$.** The set $\mathbb{Q}$ is also a field: the rational numbers are closed under addition, subtraction, multiplication, and division by numbers different from 0. The field axioms hold with the usual rules of arithmetic.

- **The real numbers $\mathbb{R}$.** The real numbers form the fundamental field of analysis. They satisfy all field axioms and, in addition, possess an order and the completeness property, both of which are essential for analysis.

- **The complex numbers $\mathbb{C}$.** The set $\mathbb{C}$ is also a field. Every nonzero complex element has a multiplicative inverse, and the usual rules of arithmetic hold as well.

- **The smallest field $\mathbb{F}_2$.** The field $\mathbb{F}_2$ consists of exactly two elements, namely 0 and 1. Addition and multiplication are defined modulo 2. Thus

  $$1 + 1 = 0, \qquad 1 \cdot 1 = 1.$$

  This field is the smallest possible field.

These examples show that the field axioms describe a very general algebraic structure. In analysis, the real numbers are usually at the center of attention, but many arguments can be transferred to other fields.

Chapter 4

The Order Axioms

So far, we have first considered the real numbers as a *field*. This already allows us to calculate in a structured way: additions, subtractions, multiplications, and divisions (by nonzero numbers) are possible, and the usual laws of arithmetic are available. For analysis, however, this is not yet sufficient.

Indeed, in many central questions of analysis, it is not enough to study equations. We also want to formulate statements of the form

$$x > 0, \qquad x < y, \qquad x \leq y$$

precisely and work with them reliably. Only in this way do estimates, monotonicity, convergence, interval considerations, and many other basic ideas of analysis become possible. In particular, we need an order on $\mathbb{R}$ that is compatible with the algebraic structure.

In the following, we therefore introduce the *order axioms*. They make $\mathbb{R}$ not merely a field, but an *ordered field*. The idea is to distinguish a set of *positive* numbers from which the entire order is then obtained.

4.1 The Order Axioms

Definition 65 (Positive Elements). *In $\mathbb{R}$, let a subset of the numbers be distinguished as the set of* positive elements. *For these positive elements, we require the following three axioms:*

1. ***Trichotomy:*** *For every $x \in \mathbb{R}$, exactly one of the three statements*

$$x > 0, \qquad -x > 0, \qquad x = 0.$$

holds.

2. ***Closure under addition:*** *If $x > 0$ and $y > 0$, then also*

$$x + y > 0.$$

holds.

3. ***Closure under multiplication:*** *If $x > 0$ and $y > 0$, then also*

$$xy > 0.$$

holds.

These axioms describe the order on $\mathbb{R}$ in a way that is compatible with arithmetic in a field. In particular, the property of positivity is preserved under addition and multiplication. This is precisely what is crucial for later estimates.

Definition 66 (Comparison symbols). *On the basis of the positive numbers, for $x, y \in \mathbb{R}$ we define the comparison relations by*

$$x > y \;:\Longleftrightarrow\; x - y > 0,$$

$$x < y \;:\Longleftrightarrow\; y > x,$$

$$x \geq y \;:\Longleftrightarrow\; (x > y) \text{ or } (x = y),$$

$$x \leq y \;:\Longleftrightarrow\; (x < y) \text{ or } (x = y).$$

With these definitions, the order is completely reduced to differences: in order to decide whether x is greater than y, one only has to consider the difference $x - y$. Thus, in many cases, inequalities can be reduced to statements about positive numbers.

Remark 67. *The order axioms are chosen so that they harmonize with the already known field axioms. They imply numerous elementary rules for calculating with inequalities, for example that one may add the same number to both sides of an inequality, or that an inequality is preserved when multiplied by a positive number. We shall derive these rules systematically in this chapter.*

4.2 Consequences

4.2.1 Comparability and Extremal Values

Let $a, b, x, y, z \in \mathbb{R}$.

Proposition 68 (Comparability of real numbers). *For all $x, y \in \mathbb{R}$, exactly one of the following three statements holds:*

$$x < y, \qquad x = y, \qquad x > y.$$

Proof. We consider the difference $x - y \in \mathbb{R}$. By the trichotomy axiom, exactly one of the following three statements holds:

$$x - y > 0, \qquad -(x - y) > 0, \qquad x - y = 0.$$

In the first case, by definition, $x > y$, since $x - y > 0$ means precisely $x > y$.

In the second case, we have

$$-(x - y) = y - x > 0,$$

so $y > x$, that is, $x < y$.

In the third case, $x - y = 0$, hence $x = y$.

Thus it has been shown that at least one of the three statements holds. That no two different ones can hold simultaneously follows directly from the axioms of order: if, for example, $x > y$ and $x = y$, then we would have both $x - y > 0$ and $x - y = 0$, which is impossible. Similarly, $x > y$ and $x < y$ exclude each other, because then both $x - y > 0$ and $y - x > 0$ would hold, and hence, by closure under addition,

$$(x - y) + (y - x) = 0 > 0,$$

contradicting $0 \not> 0$. □

Based on this comparability, we can define the maximum and the minimum of any two real numbers.

Definition 69 (Maximum and minimum). *For $x, y \in \mathbb{R}$, we define*

$$\max(x, y) := \begin{cases} x, & \text{if } x \geq y, \\ y, & \text{if } x < y, \end{cases}$$

$$\min(x, y) := \begin{cases} x, & \text{if } x \leq y, \\ y, & \text{if } x > y. \end{cases}$$

This definition is well-defined, because by the preceding proposition exactly one of the cases $x < y$, $x = y$, or $x > y$ holds for x and y.

Lemma 70 (Characterization of maximum and minimum). *For all $x, y \in \mathbb{R}$, we have*

$$\max(x, y) \geq x, \qquad \max(x, y) \geq y,$$

and

$$\min(x, y) \leq x, \qquad \min(x, y) \leq y.$$

Moreover,

$$\min(x, y) \leq \max(x, y).$$

Proof. We distinguish two cases.

Case 1: $x \geq y$. Then, by definition, $\max(x, y) = x$ and $\min(x, y) = y$. Hence

$$\max(x, y) = x \geq x, \qquad \max(x, y) = x \geq y,$$

and also

$$\min(x, y) = y \leq x, \qquad \min(x, y) = y \leq y.$$

Furthermore, $y \leq x$, and therefore $\min(x, y) \leq \max(x, y)$.

Case 2: $x < y$. Then, by definition, $\max(x, y) = y$ and $\min(x, y) = x$. It follows that

$$\max(x, y) = y \geq x, \qquad \max(x, y) = y \geq y,$$

and also

$$\min(x, y) = x \leq x, \qquad \min(x, y) = x \leq y.$$

Thus, in this case as well, $\min(x, y) \leq \max(x, y)$.

Since the two cases exhaust all possibilities, the claim is proved. □

Remark 71. *The maximum and minimum of two numbers are fundamental tools for estimates. Many inequalities can later be conveniently written in the form*

$$\min(x, y) \leq z \leq \max(x, y)$$

These concepts are also needed repeatedly for intervals, bounds, and convergence arguments.

4.2.2 Transitivity of the strict order

Proposition 72 (Transitivity of $<$). *For all $x, y, z \in \mathbb{R}$, the following implication holds:*

$$x < y \text{ and } y < z \implies x < z.$$

Proof. By the definition of the order relation, we have

$$x < y \iff y - x > 0 \quad \text{and} \quad y < z \iff z - y > 0.$$

Thus, the assumptions imply

$$y - x > 0 \quad \text{and} \quad z - y > 0.$$

Using the addition axiom for positive numbers, we obtain

$$(y - x) + (z - y) > 0.$$

Since the terms on the left-hand side combine to

$$(y - x) + (z - y) = z - x.$$

it follows that

$$z - x > 0,$$

and therefore, by the definition of the order relation,

$$x < z.$$

This proves the transitivity of the strict order. □

4.2.3 Translation Invariance of the Order

Proposition 73 (Translation invariance). *Let* $a, x, y \in \mathbb{R}$.

1. *If* $x < y$, *then* $a + x < a + y$.
2. *If* $x \leq y$, *then* $a + x \leq a + y$.

Proof. We first prove the strict version.

Assume that $x < y$. Then, by definition,

$$y - x > 0.$$

Now consider the difference of the translated numbers:

$$(a + y) - (a + x) = y - x.$$

Thus, since

$$(a + y) - (a + x) > 0$$

it follows by the definition of the order that

$$a + x < a + y.$$

We now turn to the non-strict version. If $x \leq y$, then exactly two cases may occur:

Case 1: $x < y$. Then, by the part already proved,

$$a + x < a + y,$$

and hence in particular

$$a + x \leq a + y.$$

Case 2: $x = y$. Then it follows immediately that

$$a + x = a + y,$$

and therefore also

$$a + x \leq a + y.$$

In both cases, therefore, $a + x \leq a + y$. This proves translation invariance. □

Remark 74. *The strict inequality is preserved when the same number is added to both sides. For the non-strict inequality, the same argument suffices, with the equality case taken into account separately.*

4.2.4 Change of Sign and Reversal of Order

Proposition 75 (Change of sign). *For all $x, y \in \mathbb{R}$, one has*

$$x < y \quad \Longleftrightarrow \quad -x > -y.$$

Proof. By definition,

$$x < y \Longleftrightarrow y - x > 0.$$

On the other hand,

$$(-x) - (-y) = y - x.$$

Therefore,

$$y - x > 0 \Longleftrightarrow (-x) - (-y) > 0.$$

Using the definition of the order, this gives precisely

$$x < y \iff -x > -y.$$

This proves the assertion. □

4.2.5 Additive Compatibility of the Order

Proposition 76 (Additive compatibility). *For all $x, y, a, b \in \mathbb{R}$, the following implication holds:*

$$x < y \text{ and } a < b \quad \Longrightarrow \quad x + a < y + b.$$

Proof. From $x < y$, the translation invariance of the order implies

$$a + x < a + y.$$

Moreover, from $a < b$, again by translation invariance, this time by adding y, it follows that

$$y + a < y + b.$$

Since addition in $\mathbb{R}$ is commutative, we have

$$a + y = y + a.$$

Thus we obtain

$$a + x < a + y = y + a < y + b.$$

By transitivity of the strict order, it follows that

$$a + x < y + b.$$

By commutativity of addition, this is equivalent to

$$x + a < y + b.$$

This proves the assertion. □

Remark 77. *The last statement is a typical example of an estimate in which two inequalities are combined. Such arguments are frequently used in analysis, for example in error estimates, limits, and the theory of series.*

4.2.6 Monotonicity of Multiplication by Positive Numbers

Proposition 78. *Let $x, y, a \in \mathbb{R}$. If*

$$x < y \quad \textit{and} \quad a > 0,$$

then

$$ax < ay.$$

Proof. By definition, $x < y$ implies

$$y - x > 0.$$

Since $a > 0$ is also assumed, the multiplication axiom for positive numbers yields

$$a(y - x) > 0.$$

By the distributive law, we obtain

$$a(y - x) = ay - ax.$$

Hence

$$ay - ax > 0.$$

By the definition of the order relation, this is equivalent to

$$ax < ay.$$

This proves the assertion. □

Remark 79. *The important point here is that multiplication by a* positive *number preserves the direction of an inequality. For negative factors, the inequality will later be reversed.*

4.2.7 Comparison of Products

Proposition 80. *Let $x, y, a, b \in \mathbb{R}$. If*

$$0 \leq x < y \quad \text{and} \quad 0 \leq a < b,$$

then

$$ax < by.$$

Proof. We first distinguish cases according to the sign of a.

Case 1: $a = 0$.
Then

$$ax = 0.$$

Moreover, from $0 \leq a < b$ together with $a = 0$, it follows that $b > 0$. From $0 \leq x < y$ we first obtain $y > 0$: if

$x = 0$, this is immediate, since then $0 < y$. If, on the other hand, $x > 0$, then $x < y$ and transitivity imply

$$0 < x < y,$$

so in particular $y > 0$.

Thus both $b > 0$ and $y > 0$. By the preceding result on multiplication by positive numbers, we obtain

$$0 = ax < by,$$

and hence $ax < by$.

Case 2: $a > 0$.

Then $x < y$ and the preceding result imply

$$ax < ay.$$

Since $y > 0$, the preceding result applied to $a < b$ also yields

$$ay < by.$$

By transitivity of $<$, it follows that

$$ax < by.$$

Since the desired inequality has been shown in both cases, the claim is proved. □

Remark 81. *This statement is particularly useful because it breaks a typical estimate into two steps: first one compares using the smaller factor a or b, and then using the larger number x or y. Arguments of exactly this kind are frequently needed in analysis.*

4.2.8 Monotonicity of Multiplication by Negative Numbers

Proposition 82. *Let $x, y, a \in \mathbb{R}$. If*

$$x < y \quad \textit{and} \quad a < 0,$$

then

$$ax > ay.$$

Proof. From $a < 0$, it follows by the definition of the order that

$$-a > 0.$$

From the assumption $x < y$ and the previously proved monotonicity of multiplication by positive numbers, we obtain

$$(-a)x < (-a)y.$$

Since $(-a)x = -ax$ and $(-a)y = -ay$, this gives

$$-ax < -ay.$$

Applying the already proved rule for changing signs, it follows that

$$ax > ay.$$

Thus the claim is proved. □

Remark 83. *Multiplication by a negative number therefore reverses the direction of an inequality. This is one of the most important rules for working with inequalities.*

4.2.9 Positivity of Squares and of 1

Proposition 84. *For every $x \in \mathbb{R}$ with $x \neq 0$, we have*

$$x^2 > 0.$$

In particular,

$$1 > 0.$$

Proof. First let $x \neq 0$.

By the trichotomy axiom, either $x > 0$ or $x < 0$.

Case 1: $x > 0$.

Then both factors in

$$x \cdot x = x^2$$

are positive. Hence, by closure under multiplication,

$$x^2 > 0.$$

Case 2: $x < 0$.

Then, after changing signs,

$$-x > 0.$$

Thus, by closure under multiplication,

$$(-x)(-x) > 0.$$

Since

$$(-x)(-x) = x^2$$

it follows again that

$$x^2 > 0.$$

This shows that the inequality $x^2 > 0$ holds for every $x \neq 0$.

For $x = 1$, we obtain in particular

$$1^2 > 0.$$

Since $1^2 = 1$, it follows that

$$1 > 0.$$

□

Remark 85. *The positivity of squares plays a central role in analysis. It is the basis of many estimates, for instance in the study of absolute values, distances, and convergence.*

4.2.10 Positivity of the Multiplicative Inverse

Proposition 86. *For every $x \in \mathbb{R} \setminus \{0\}$, one has*

$$x > 0 \iff x^{-1} > 0.$$

Proof. We prove the two implications separately.

"⇒": Let $x > 0$. We consider x^{-1} and, by the trichotomy axiom, distinguish the three cases

$$x^{-1} > 0, \qquad x^{-1} = 0, \qquad x^{-1} < 0.$$

The case $x^{-1} = 0$ is impossible, since $x \neq 0$, so x^{-1} exists as the inverse of x and satisfies $xx^{-1} = 1$.

Now suppose that $x^{-1} < 0$. Since $x > 0$, the monotonicity of multiplication by positive numbers, applied to the inequality

$$x^{-1} < 0,$$

implies that

$$x \cdot x^{-1} < x \cdot 0 = 0.$$

On the other hand,

$$x \cdot x^{-1} = 1.$$

Thus we would obtain $1 < 0$, contradicting $1 > 0$. Hence $x^{-1} < 0$ cannot hold.

Since neither $x^{-1} = 0$ nor $x^{-1} < 0$ is possible, the trichotomy axiom leaves only

$$x^{-1} > 0.$$

"$\Leftarrow$": Now let $x^{-1} > 0$. Applying the implication already proved to the number x^{-1}, we obtain

$$(x^{-1})^{-1} > 0.$$

Since $(x^{-1})^{-1} = x$, it follows that

$$x > 0.$$

This proves that

$$x > 0 \iff x^{-1} > 0.$$

□

Remark 87. *This statement is particularly important: positive numbers again have positive inverses. As a result, many estimates remain valid under division.*

4.2.11 Monotonicity of Reciprocals

Proposition 88. *If*

$$0 < x < y,$$

then

$$x^{-1} > y^{-1}.$$

Proof. From $0 < x < y$ it follows in particular that

$$x > 0 \quad \text{and} \quad y > 0.$$

By the preceding proposition, we therefore obtain

$$x^{-1} > 0 \quad \text{and} \quad y^{-1} > 0.$$

Hence also

$$x^{-1}y^{-1} > 0.$$

We now multiply the inequality $x < y$ by the positive number $x^{-1}y^{-1}$. By the monotonicity of multiplication by positive numbers, it follows that

$$x(x^{-1}y^{-1}) < y(x^{-1}y^{-1}).$$

Using associativity and commutativity of multiplication, together with the identities $xx^{-1} = 1$ and $yy^{-1} = 1$, we obtain

$$x(x^{-1}y^{-1}) = (xx^{-1})y^{-1} = 1 \cdot y^{-1} = y^{-1},$$

and similarly

$$y(x^{-1}y^{-1}) = x^{-1}(yy^{-1}) = x^{-1} \cdot 1 = x^{-1}.$$

Thus

$$y^{-1} < x^{-1},$$

or equivalently

$$x^{-1} > y^{-1}.$$

This proves the claim. □

Remark 89. *The function $x \mapsto x^{-1}$ is strictly decreasing on $(0, \infty)$. This property will be needed repeatedly later when working with fractions, estimates, and limits.*

4.2.12 Ordered Fields

Definition 90 (Ordered field)**.** *A field K is called* ordered *if there is a set of positive elements in K such that the order axioms hold; in particular:*

1. *For every $x \in K$, exactly one of the statements $x > 0$, $-x > 0$, or $x = 0$ holds.*
2. *If $x > 0$ and $y > 0$, then $x + y > 0$.*
3. *If $x > 0$ and $y > 0$, then $xy > 0$.*

The real numbers $\mathbb{R}$ form an ordered field with their usual order. The rational numbers $\mathbb{Q}$ are also ordered, since the ordinary order on $\mathbb{Q}$ satisfies the same properties as the order on $\mathbb{R}$.

Remark 91. *For $\mathbb{Q}$ and $\mathbb{R}$, the order is therefore not merely present; it is compatible with arithmetic in the usual way. This is precisely what makes these two fields so important for analysis.*

Proposition 92. *The field $\mathbb{F}_2$ of residue classes modulo 2 cannot be ordered.*

Proof. Suppose that $\mathbb{F}_2$ were ordered. Then, by the previously proved consequence concerning squares, we would in particular have

$$1 > 0,$$

since $1 = 1^2$ and $1 \neq 0$.

However, in $\mathbb{F}_2$ we have

$$1 + 1 = 0$$

and therefore closure under addition of positive elements would imply

$$1 > 0 \text{ and } 1 > 0 \quad \Longrightarrow \quad 1 + 1 > 0.$$

Thus we would obtain

$$0 > 0,$$

which contradicts the trichotomy axiom. Hence $\mathbb{F}_2$ cannot be ordered. □

Proposition 93. *The field $\mathbb{C}$ of complex numbers also cannot be ordered.*

Proof. Suppose that $\mathbb{C}$ were ordered. Then, by the previously proved consequence concerning squares, every element different from 0 would satisfy

$$z \neq 0 \implies z^2 > 0.$$

In particular, since $i \neq 0$, we would have to have

$$i^2 > 0$$

. However,

$$i^2 = -1$$

so it follows that

$$-1 > 0.$$

On the other hand, we also have

$$1 > 0,$$

because $1 = 1^2$ and $1 \neq 0$.

Closure under addition of positive elements would then imply

$$1 > 0 \text{ and } -1 > 0 \implies 1 + (-1) > 0,$$

that is,

$$0 > 0,$$

a contradiction. Therefore $\mathbb{C}$ cannot be ordered. □

Remark 94. *The impossibility of ordering $\mathbb{C}$ is not merely a technical limitation; it reveals a fundamental difference between real and complex numbers: in the complex numbers there are elements whose square is negative, and this is incompatible with an order that is compatible with arithmetic.*

4.3 The Natural Numbers as a Subset of the Real Numbers

In the preceding sections, we first encountered the real numbers as an ordered field. The order axioms ensure that the order is compatible with the arithmetic operations and thereby exclude many contradictory phenomena that may occur in other algebraic structures. For example, in $\mathbb{F}_2$ the equation $1+1=0$ is unavoidable, whereas in $\mathbb{C}$ the element i has the property $i^2=-1$. Such situations are not compatible with an order in which positive numbers remain stable under addition and multiplication.

In $\mathbb{R}$, by contrast, we can recover the natural numbers in a particularly simple way: they arise by repeatedly adding 1 to 0. We now want to make this idea precise. To this end, we consider a subset of the real numbers which, starting from 0, is closed under the operation $x \mapsto x+1$. In this sense, $\mathbb{N}$ may be regarded as a set generated within $\mathbb{R}$.

4.3.1 The Peano Axioms

Definition 95 (The set generated by 0 and the successor). *Let $A \subset \mathbb{R}$ be the smallest subset with the following two properties:*

1. $0 \in A$,
2. *if* $x \in A$, *then also* $x+1 \in A$.

Then A consists precisely of those real numbers that can be reached from 0 by successive addition of 1.

Since $x + 1 \in A$ holds for every $x \in A$, the following map naturally serves as the successor function:

$$n\colon A \to A, \qquad n(x) := x + 1.$$

Intuitively, $n(x)$ is therefore the value that follows x.

The set of natural numbers can now be characterized by the distinguished element 0 and the successor function n. The essential structure is described by the Peano axioms.

Definition 96 (Peano axioms). *For the set A with the distinguished element 0 and the successor function $n\colon A \to A$, the Peano axioms are as follows:*

P1 Injectivity of the successor:

$$x \neq y \implies n(x) \neq n(y).$$

Thus different elements of A have different successors.

P2 No element has 0 as its successor:

$$0 \notin n(A).$$

Hence there is no $x \in A$ with $n(x) = 0$.

P3 Induction axiom: *If $M \subset A$ is a subset with*

$$0 \in M \quad \textit{and} \quad x \in M \Rightarrow n(x) \in M,$$

then

$$M = A.$$

These axioms express exactly the structure one expects of the natural numbers: starting at 0, forming successors, having no "backward resolution" to the zero point, and finally the principle of induction as a characterization of all subsets that begin with 0 and are closed under the successor operation.

In the next step, we shall verify these axioms for the set A defined in this way. Only then does it become clear that the natural numbers can indeed be regarded as a structure contained in the real numbers.

4.3.2 Verification of the Peano Axioms

We now show that the previously defined set $A \subset \mathbb{R}$, together with the distinguished element 0 and the successor function

$$n \colon A \to A, \qquad n(x) = x + 1,$$

does indeed satisfy the Peano axioms. Thus we may identify A with the set of natural numbers.

Proposition 97. *The set A satisfies the Peano axioms **P1**–**P3**.*

Proof. We verify the three axioms one by one.

Axiom P1: Let $x, y \in A$ with

$$n(x) = n(y).$$

By the definition of the successor function, this means

$$x + 1 = y + 1.$$

Since $\mathbb{R}$ is a field, we may subtract 1 from both sides. It follows that

$$x = y.$$

Hence we have shown that

$$x \neq y \Rightarrow n(x) \neq n(y).$$

Thus **P1** holds.

Axiom P3: This axiom is already contained in the definition of A. Indeed, the set A was defined precisely as the smallest subset of $\mathbb{R}$ that contains

$$0 \in A$$

and that, together with each element x, also contains its successor $x + 1$.

Thus, if $M \subset A$ is a subset with

$$0 \in M \quad \text{and} \quad x \in M \Rightarrow n(x) \in M,$$

then, by the minimality of A, we must already have $A \subset M$. Since conversely $M \subset A$ is assumed anyway, it follows that

$$M = A.$$

Hence **P3** is satisfied.

Axiom P2: First we show that

$$x \geq 0 \qquad \text{for all } x \in A.$$

For this purpose, we define the subset

$$M := \{x \in A : x \geq 0\}.$$

We now show that M satisfies the hypotheses of **P3**.

Step 1: $0 \in M$.
This is clear, since $0 \geq 0$ always holds.

Step 2: From $x \in M$ it follows that $n(x) \in M$.
So let $x \in M$. Then $x \geq 0$. Since $1 > 0$, in particular

$$0 < 1.$$

By the translation invariance of the order, we obtain

$$x < x + 1 = n(x).$$

Since moreover $x \geq 0$, transitivity yields

$$0 \leq n(x).$$

Thus $n(x) \in M$.

Since M satisfies both hypotheses of **P3**, it follows that

$$M = A.$$

Hence we have shown that

$$x \geq 0 \qquad \text{for all } x \in A.$$

We now verify **P2**. Suppose that **P2** were false. Then there would be an $x \in A$ with

$$0 = n(x) = x + 1.$$

Subtracting 1 gives

$$x = -1.$$

On the other hand, we have just shown that for all elements of A

$$x \geq 0$$

holds. Hence in particular one would have to have $-1 \geq 0$.

However, this contradicts the already established positivity of 1 and the change of sign: from $1 > 0$ it follows that

$$-1 < 0.$$

This gives a contradiction. Consequently, no $x \in A$ can satisfy the equation $n(x) = 0$. Thus

$$0 \notin n(A).$$

This proves **P2** as well.

Therefore A, 0, and n satisfy all Peano axioms. □

Theorem 98 (Uniqueness of the natural numbers up to isomorphism). *Let*

$$(A, 0_A, n_A) \quad \textit{and} \quad (B, 0_B, n_B)$$

be two structures satisfying the Peano axioms. Then there exists exactly one bijective map

$$\varphi\colon A \to B$$

such that

$$\varphi(0_A) = 0_B$$

and

$$\varphi(n_A(x)) = n_B(\varphi(x)) \qquad \text{for all } x \in A.$$

In particular, all models of the Peano axioms are uniquely determined up to isomorphism.

Proof. The idea is to assign to each element of A a uniquely determined element of B by recursion.

We first define a subset

$$M \subset A$$

by declaring that

$$x \in M$$

if and only if there exists a uniquely determined element $\varphi(x) \in B$ satisfying

$$\varphi(0_A) = 0_B$$

and

$$\varphi(n_A(x)) = n_B(\varphi(x)).$$

We now use the induction axiom to show that $M = A$.

1. Step: $0_A \in M$.
For 0_A, the value is uniquely prescribed by

$$\varphi(0_A) := 0_B.$$

2. Step: From $x \in M$ it follows that $n_A(x) \in M$.

If $\varphi(x)$ has already been uniquely determined, we define

$$\varphi(n_A(x)) := n_B(\varphi(x)).$$

Thus the value at the successor is also uniquely determined.

By Peano axiom **P3**, it follows that

$$M = A.$$

Hence φ is recursively defined on all of A.

We now show that φ is injective.

Let $x, y \in A$ with

$$\varphi(x) = \varphi(y).$$

Consider the set

$$N := \{x \in A : \forall y \in A,\ \varphi(x) = \varphi(y) \Rightarrow x = y\}.$$

By the injectivity of the successor maps (**P1**), we have:

$$\varphi(n_A(x)) = n_B(\varphi(x)) = n_B(\varphi(y)) = \varphi(n_A(y))$$

implies

$$n_A(x) = n_A(y),$$

and therefore, again by **P1**,

$$x = y.$$

By complete induction, it follows that φ is injective.

For surjectivity, consider the set

$$S := \varphi(A) \subset B.$$

We have

$$0_B = \varphi(0_A) \in S.$$

Moreover, from $b = \varphi(x) \in S$ it follows that

$$n_B(b) = n_B(\varphi(x)) = \varphi(n_A(x)) \in S.$$

Thus S satisfies the hypotheses of the induction axiom in B. Hence

$$S = B.$$

Thus φ is surjective.

Therefore φ is bijective and compatible with the successor maps. Hence the two structures are isomorphic. □

Remark 99. *This result shows that the Peano axioms completely characterize the natural numbers: every structure satisfying these axioms has, up to a renaming of its elements, the same structure as the natural numbers.*

Therefore, we may identify the previously constructed set A with the usual set of natural numbers and henceforth write

$$A = \mathbb{N}.$$

Thus we see that the natural numbers are contained in every ordered field.

Definition 100. *The set A is called the set of* natural numbers *and is denoted by*

$$\mathbb{N}$$

We have therefore constructed the natural numbers as a structure defined within the real numbers.

Remark 101 (The induction axiom and complete induction). *The Peano axiom* ***P3*** *is precisely the structural foundation of the principle of complete induction, which we already encountered in the first chapter.*

To see this, let

$$A(n)$$

be a statement about natural numbers $n \in \mathbb{N}$. We consider the set

$$M := \{n \in \mathbb{N} : A(n) \text{ is true}\}.$$

The two assumptions of the Peano axiom ***P3*** *then correspond exactly to the two steps of a proof by induction:*

1. ***Base case:***

   $$A(0) \text{ is true.}$$

   This means precisely that

   $$0 \in M.$$

2. ***Induction step:***

   From the truth of $A(n)$, the truth of $A(n+1)$ follows.

Because of the successor map

$$n(x) = x + 1$$

this means precisely that

$$n \in M \Rightarrow n + 1 = n(n) \in M.$$

If both conditions are satisfied, then the Peano axiom ***P3*** *yields*

$$M = \mathbb{N}.$$

Thus the statement $A(n)$ is true for all natural numbers n.

The principle of complete induction is therefore not an additional property of the natural numbers, but follows directly from their characterization by the Peano axioms.

4.4 The Absolute Value

The *absolute value* of a real number measures its distance from 0. It is therefore a central tool in analysis whenever one is concerned not only with exact equalities, but especially with estimates.

Definition 102 (Absolute value). *For $x \in \mathbb{R}$, we define the absolute value by*

$$|x| := \begin{cases} x, & \textit{if } x \geq 0, \\ -x, & \textit{if } x < 0. \end{cases}$$

Equivalently, one has

$$|x| = \max(x, -x).$$

The second form immediately shows that $|x|$ is always nonnegative and at the same time provides an upper bound for both x and $-x$.

Proposition 103 (Basic properties of the absolute value). *For all $x, y \in \mathbb{R}$, the following hold:*

1. $|x| \geq 0$ *and* $|x| = 0 \Leftrightarrow x = 0$.

2. $|xy| = |x|\,|y|$.

3. $|x + y| \leq |x| + |y|$.

Proof. We prove the three statements separately.

Ad 1. By definition, either $|x| = x$ with $x \geq 0$, or $|x| = -x$ with $x < 0$. In both cases, therefore,

$$|x| \geq 0.$$

Moreover, from $|x| = 0$ we first obtain:

- If $x \geq 0$, then $|x| = x$, hence $x = 0$.

- If $x < 0$, then $|x| = -x = 0$, hence again $x = 0$.

Conversely, if $x = 0$, then immediately $|x| = 0$. Thus, altogether,

$$|x| = 0 \Longleftrightarrow x = 0.$$

Ad 2. We distinguish cases according to the signs of x and y.

- If $x \geq 0$ and $y \geq 0$, then

$$|x| = x, \qquad |y| = y,$$

and hence

$$|xy| = xy = |x|\,|y|.$$

- If $x \geq 0$ and $y < 0$, then

$$|x| = x, \qquad |y| = -y.$$

Since then $xy < 0$, we have

$$|xy| = -(xy) = x(-y) = |x|\,|y|.$$

- If $x < 0$ and $y \geq 0$, then analogously

$$|xy| = |x|\,|y|.$$

- If $x < 0$ and $y < 0$, then

$$|x| = -x, \qquad |y| = -y,$$

and since $xy > 0$, we have

$$|xy| = xy = (-x)(-y) = |x|\,|y|.$$

Thus, in all cases,

$$|xy| = |x|\,|y|.$$

Ad 3. First, we observe that for every $x \in \mathbb{R}$,

$$x \leq |x| \qquad \text{and} \qquad -x \leq |x|$$

holds. This follows immediately from the definition $|x| = \max(x, -x)$.

Therefore, by adding the corresponding inequalities, we obtain

$$x + y \leq |x| + |y| \qquad \text{and} \qquad -(x+y) \leq |x| + |y|.$$

Now

$$|x + y| = \max(x + y,\ -(x + y)).$$

Since both numbers on the right-hand side are at most $|x| + |y|$, it follows that

$$|x + y| \leq |x| + |y|.$$

This proves the triangle inequality. □

Remark 104. *The third property is the so-called* triangle inequality. *It is one of the most important inequalities in analysis. It makes it possible to control series, sequences, limits, and error estimates.*

Remark 105. *The properties of the absolute value do not necessarily depend on the underlying field being ordered. The field of complex numbers $\mathbb{C}$ provides an important example of this.*

Although $\mathbb{C}$ cannot be an ordered field, it nevertheless has an absolute value with the properties

$$|z| \geq 0, \qquad |z| = 0 \iff z = 0,$$

$$|zw| = |z|\,|w|,$$

as well as the triangle inequality

$$|z + w| \leq |z| + |w| \qquad \text{for all } z, w \in \mathbb{C}.$$

For a complex number

$$z = a + bi$$

the complex absolute value is defined by

$$|z| = \sqrt{a^2 + b^2}.$$

Geometrically, it describes the distance of the point (a, b) *from the origin of the plane.*

Thus $\mathbb{C}$ *shows that the fundamental properties of the absolute value can also be meaningfully defined on non-ordered fields. In* $\mathbb{R}$*, the absolute value naturally arises from the order, whereas in* $\mathbb{C}$ *it arises from the geometry of the plane.*

4.4.1 Further Properties of the Absolute Value

Proposition 106. *For all* $x, y \in \mathbb{R}$ *with* $y \neq 0$*, the following statements hold:*

1. $|1| = 1$ *and* $|-1| = 1$.
2. $|-x| = |x|$.
3. $\left|\dfrac{x}{y}\right| = \dfrac{|x|}{|y|}$.

4. $|x - y| \geq |x| - |y|$.

5. $|x + y| \geq |x| - |y|$.

Proof. We prove the statements one after another.

For 1. Since $1 > 0$, the definition of the absolute value gives

$$|1| = 1.$$

Moreover, $-1 < 0$, hence

$$|-1| = -(-1) = 1.$$

For 2. Let $x \in \mathbb{R}$. We distinguish two cases.

If $x \geq 0$, then $-x \leq 0$. If $x = 0$, the assertion is trivial. If $x > 0$, then

$$|-x| = -(-x) = x = |x|.$$

If $x < 0$, then $-x > 0$, and therefore

$$|-x| = -x = |x|.$$

In both cases, it follows that

$$|-x| = |x|.$$

For 3. Let $y \neq 0$. Then also $|y| \neq 0$, since $|y| = 0$ would imply $y = 0$ by the preceding basic fact.

Using the multiplicativity of the absolute value, we obtain

$$|y| \left|y^{-1}\right| = |y \cdot y^{-1}| = |1| = 1.$$

Thus

$$\left|y^{-1}\right| = \frac{1}{|y|}.$$

Now, for the quotient we have

$$\left|\frac{x}{y}\right| = |x \cdot y^{-1}| = |x|\,|y^{-1}| = |x| \cdot \frac{1}{|y|} = \frac{|x|}{|y|}.$$

For 4. From the triangle inequality, we first obtain

$$|x| = |(x-y)+y| \leq |x-y| + |y|.$$

Subtracting $|y|$ from both sides gives

$$|x| - |y| \leq |x-y|.$$

Hence

$$|x-y| \geq |x| - |y|.$$

For 5. This follows similarly from the triangle inequality:

$$|x| = |(x+y)-y| \leq |x+y| + |y|.$$

After subtracting $|y|$, we therefore get

$$|x| - |y| \leq |x+y|.$$

Thus

$$|x+y| \geq |x| - |y|$$

is shown. □

Proposition 107 (Reverse Triangle Inequality). *For all $x, y \in \mathbb{R}$, one has*

$$\big||x| - |y|\big| \leq |x - y|.$$

Proof. We use the triangle inequality twice.

First, we have

$$x = (x - y) + y.$$

Hence it follows that

$$|x| = |(x - y) + y| \leq |x - y| + |y|.$$

Thus we obtain

$$|x| - |y| \leq |x - y|.$$

Now interchanging the roles of x and y, we similarly get

$$|y| = |(y - x) + x| \leq |y - x| + |x|.$$

Since $|y - x| = |x - y|$, it follows that

$$|y| - |x| \leq |x - y|.$$

This is equivalent to

$$-(|x| - |y|) \leq |x - y|.$$

We have therefore shown both inequalities

$$|x| - |y| \leq |x - y| \qquad \text{and} \qquad -\left(|x| - |y|\right) \leq |x - y|$$

Therefore, the absolute value of $|x|-|y|$ is at most $|x-y|$, that is,

$$\big||x| - |y|\big| \leq |x - y|.$$

This proves the assertion. □

Remark 108. *The reverse triangle inequality is a very useful estimate. It says that the absolute values of two numbers can differ by no more than the numbers themselves.*

4.5 The Archimedean Axiom

In addition to the field axioms and the order axioms, another property of the real numbers plays an important role in analysis: the *Archimedean axiom.* It states that there are no real numbers that are "infinitely large" in comparison with the natural numbers. In other words, by adding a positive number sufficiently many times, one can eventually exceed any prescribed number.

Geometrically, this corresponds to the idea that on a line any given segment can be surpassed after finitely many steps by repeatedly laying off a fixed positive segment. This property is essential for the understanding of lengths and magnitudes in analysis.

Definition 109 (Archimedean axiom). *An ordered field K is called* Archimedean *if for all $x, y \in K$ with $x > 0$ there exists a natural number $n \in \mathbb{N}$ such that*

$$nx > y.$$

Equivalently: For every $y \in K$ there exists an $n \in \mathbb{N}$ such that

$$y < n.$$

Remark 110. *The first formulation expresses that, with sufficiently many summands equal to x, one can exceed any prescribed bound y. The second formulation is the special case $x = 1$.*

Proposition 111. *The ordered fields $\mathbb{Q}$ and $\mathbb{R}$ are Archimedean.*

Proof. For $\mathbb{Q}$ and $\mathbb{R}$, the Archimedean property follows from the usual order of numbers: for every real number, there is a natural number that is larger. In particular, for every y one can choose an $n \in \mathbb{N}$ such that $n > y$. Hence the general form $nx > y$ is also satisfied for every $x > 0$, since after division by x it reduces to the special form. □

Remark 112. *There are also ordered fields that are* not *Archimedean. A brief example is the field of rational functions*

$$\mathbb{R}(t),$$

if t is regarded as a positive quantity that is larger than every real number. Then t is "infinitely large" in comparison with the natural numbers, and the Archimedean axiom is violated.

Such fields play no role in the analysis of the real numbers, but they show that the Archimedean property is an additional, genuine property of ordered fields and does not already follow from the order axioms.

4.5.1 Consequences

Proposition 113 (Integer part of a real number). *For every $x \in \mathbb{R}$ there exists exactly one $n \in \mathbb{Z}$ such that*

$$n \leq x < n + 1.$$

This uniquely determined n is called the floor *or* integer part *of x, and one writes*

$$n = \lfloor x \rfloor.$$

Equivalently, there exists exactly one $m \in \mathbb{Z}$ such that

$$m - 1 < x \leq m.$$

This uniquely determined m is called the ceiling *of x and is denoted by*

$$m = \lceil x \rceil.$$

Proof. We first prove the existence of an $n \in \mathbb{Z}$ such that $n \leq x < n + 1$.

By the Archimedean axiom, there exist natural numbers $n_1, n_2 \in \mathbb{N}$ such that

$$-n_2 < x < n_1.$$

Set

$$y := x + n_2.$$

Then

$$0 < y < n_1 + n_2.$$

Now consider the set

$$S := \{k \in \mathbb{N} : y < k\}.$$

This set is nonempty, since $n_1 + n_2 \in S$. By the well-ordering principle, S has a least element; denote it by m.

Then, by the definition of S,

$$y < m.$$

If, in addition, $m - 1 > y$, then already $m - 1$ would belong to S, contradicting the minimality of m. Hence

$$m - 1 \leq y < m$$

must hold. Since $y = x + n_2$, it follows that

$$m - n_2 - 1 \leq x < m - n_2.$$

If we set

$$n := m - n_2 - 1 \in \mathbb{Z},$$

we obtain precisely

$$n \leq x < n + 1.$$

This proves existence.

For uniqueness, let $n, m \in \mathbb{Z}$ satisfy

$$n \leq x < n + 1 \qquad \text{and} \qquad m \leq x < m + 1.$$

Assume that $n < m$. Then, by the discreteness of the integers,

$$n + 1 \leq m.$$

Together with $m \leq x < n + 1$, this yields

$$n + 1 \leq m \leq x < n + 1,$$

a contradiction. Thus $n < m$ is impossible. Analogously, $m < n$ is also impossible. Consequently,

$$n = m.$$

This proves uniqueness.

The second statement follows immediately from the first, applied to $-x$: there exists exactly one $k \in \mathbb{Z}$ such that

$$k \leq -x < k + 1.$$

Multiplying by -1 and taking into account the reversal of the order gives

$$-k - 1 < x \leq -k.$$

Setting

$$m := -k,$$

we obtain

$$m - 1 < x \leq m.$$

This number is unique as well. Hence the assertion is proved. □

Proposition 114 (Archimedean approximation by reciprocals). *For every $\varepsilon > 0$ there exists an $n \in \mathbb{N}$ such that*

$$\frac{1}{n} < \varepsilon.$$

Proof. Let $\varepsilon > 0$. Then also $\varepsilon^{-1} > 0$. By the Archimedean axiom, there therefore exists an $n \in \mathbb{N}$ such that

$$n > \varepsilon^{-1}.$$

In particular, $n > 0$. Using the previously proved monotonicity of the reciprocal in positively ordered fields, it follows from

$$\varepsilon^{-1} < n$$

that

$$n^{-1} < (\varepsilon^{-1})^{-1} = \varepsilon.$$

Hence

$$\frac{1}{n} < \varepsilon.$$

□

Remark 115. *This statement is one of the most important consequences of the Archimedean axiom: no matter how small a positive bound is, it can be undercut by some reciprocal $1/n$. Precisely this property will be used repeatedly later in the study of limits and convergence.*

4.5.2 Bernoulli's Inequality

Proposition 116 (Bernoulli's inequality). *For all $x \geq -1$ and all $n \in \mathbb{N}_0$, one has*

$$(1+x)^n \geq 1+nx.$$

In particular, this also holds for all $x \geq 1$.

Proof. We prove the statement by mathematical induction on n.

Base case: For $n = 0$, we have

$$(1+x)^0 = 1$$

and at the same time

$$1 + 0 \cdot x = 1.$$

Thus

$$(1+x)^0 = 1 \geq 1 = 1 + 0 \cdot x.$$

This proves the base case.

Induction step: Now assume that for some fixed $n \in \mathbb{N}_0$ we already have

$$(1+x)^n \geq 1+nx$$

as the induction hypothesis. We then show that

$$(1+x)^{n+1} \geq 1+(n+1)x.$$

From the induction hypothesis, after multiplication by the nonnegative number $1 + x$, it follows that

$$(1+x)^{n+1} = (1+x)^n(1+x) \geq (1+nx)(1+x).$$

We are allowed to multiply by $1 + x$ because $x \geq -1$ implies

$$1 + x \geq 0$$

and therefore the direction of the inequality is preserved.

Now we expand:

$$(1+nx)(1+x) = 1+x+nx+nx^2 = 1+(n+1)x+nx^2.$$

Since $n \in \mathbb{N}_0$ and $x^2 \geq 0$, we have

$$nx^2 \geq 0.$$

Hence

$$(1+nx)(1+x) \geq 1+(n+1)x.$$

Combining these estimates gives

$$(1+x)^{n+1} \geq 1+(n+1)x.$$

This proves the induction step.

Since both the base case and the induction step have been established, the assertion is proved for all $n \in \mathbb{N}_0$.

□

Remark 117. *Bernoulli's inequality is one of the most important elementary estimates in analysis. It provides a simple lower bound for powers and will later be used frequently in connection with series, questions of convergence, and growth estimates.*

4.5.3 Growth and Decay of Powers

Proposition 118 (Powers of a number greater than 1). *Let $b > 1$. Then for every $r \in \mathbb{R}$ there exists an $n \in \mathbb{N}$ such that*

$$\exists n \in \mathbb{N}: \quad b^n > r.$$

Proof. Set

$$x := b - 1.$$

Then $x > 0$ and

$$b = 1 + x.$$

By Bernoulli's inequality, for all $n \in \mathbb{N}$ we have

$$b^n = (1 + x)^n \geq 1 + nx.$$

Now let $r \in \mathbb{R}$ be arbitrary. Since $x > 0$, the Archimedean axiom yields an $n \in \mathbb{N}$ such that

$$n > \frac{r - 1}{x}.$$

Multiplying by $x > 0$ gives

$$nx > r - 1,$$

and hence

$$1 + nx > r.$$

Together with Bernoulli's inequality, it follows that

$$b^n \geq 1 + nx > r.$$

Thus it has been shown that for every $r \in \mathbb{R}$ there exists an $n \in \mathbb{N}$ with $b^n > r$. □

Proposition 119 (Powers of a number between 0 and 1). *Let*

$$0 < b < 1.$$

Then for every $\varepsilon > 0$ there exists an $n \in \mathbb{N}$ such that

$$\exists n \in \mathbb{N} : \quad b^n < \varepsilon.$$

Proof. Set

$$c := \frac{1}{b}.$$

Since $0 < b < 1$, we have $c > 1$. By the preceding proposition, for every $\varepsilon > 0$ there exists an $n \in \mathbb{N}$ such that

$$c^n > \frac{1}{\varepsilon}.$$

Since $c > 0$, we may apply the monotonicity of reciprocals. From

$$c^n > \frac{1}{\varepsilon}$$

it follows that

$$\frac{1}{c^n} < \varepsilon.$$

Moreover,

$$\frac{1}{c^n} = \left(\frac{1}{c}\right)^n = b^n.$$

Therefore we obtain

$$b^n < \varepsilon.$$

This proves the assertion. □

Remark 120. *The two statements describe the asymptotic behavior of powers: if the base is greater than* 1*, then its powers become arbitrarily large. If the base lies strictly between* 0 *and* 1*, then its powers become arbitrarily small. These two basic facts will be used repeatedly later in the study of limits, exponential functions, and series.*

4.6 Exercises and Solutions

The following exercises are intended to strengthen the confident use of the field axioms and of the rules of calculation that follow from them. In the solutions, care is taken to explicitly name the axioms and transformations used, so that the idea of the proof remains clearly understandable.

1. Uniqueness of Neutral and Inverse Elements

1. Show, using only the field axioms, that the additive neutral element is unique.

2. Show that for every $x \in \mathbb{R}$ the additive inverse is uniquely determined.

Solution:

(1) Let 0 and $0'$ be two additive neutral elements. Then, for every $x \in \mathbb{R}$, both

$$x + 0 = x \qquad \text{and} \qquad x + 0' = x.$$

If we set $x = 0$ in the second equation, we obtain

$$0 + 0' = 0.$$

On the other hand, by the neutrality of 0, we also have

$$0 + 0' = 0'.$$

Thus it follows that

$$0 = 0'.$$

Hence the additive neutral element is uniquely determined.

(2) Let $x \in \mathbb{R}$ be fixed, and let $y, z \in \mathbb{R}$ be two additive inverses of x. Then

$$x + y = 0 \qquad \text{and} \qquad x + z = 0.$$

We now add z to both sides of the first equation. This gives

$$(x + y) + z = 0 + z.$$

By associativity and commutativity, it follows that

$$x + (y + z) = z.$$

Since $x + z = 0$, we may also write

$$(x + z) + y = 0 + y.$$

Again using associativity and commutativity, this implies

$$x + (z + y) = y.$$

Since $y + z = z + y$, we obtain altogether

$$x + (y + z) = z \qquad \text{and} \qquad x + (y + z) = y.$$

Therefore $y = z$. Thus the additive inverse is unique.

2. Standard transformations using the field axioms

1. Show that $-(x+y) = -x - y$.

2. Show that $-(-x) = x$.

3. Show that $x \cdot 0 = 0$.

Solution:

(1) We show that $(-x)+(-y)$ is an additive inverse of $x+y$. Indeed, we compute:

$$(x+y)+((-x)+(-y)) = (x+(-x))+(y+(-y)) = 0+0 = 0.$$

Thus $(-x)+(-y)$ is an additive inverse of $x+y$. By the uniqueness of the additive inverse, it follows that

$$-(x+y) = (-x)+(-y).$$

Usually, this is written briefly as

$$-(x+y) = -x - y.$$

(2) By definition, $-x$ is the additive inverse of x, hence

$$x+(-x) = 0.$$

By commutativity of addition, we also have

$$(-x)+x = 0.$$

Therefore x is an additive inverse of $-x$. Since additive inverses are unique, it follows that

$$-(-x) = x.$$

(3) Using the distributive law, we obtain

$$x \cdot 0 = x \cdot (0 + 0) = x \cdot 0 + x \cdot 0.$$

Now we add the additive inverse of $x \cdot 0$ to both sides. This gives

$$0 = x \cdot 0.$$

Hence

$$x \cdot 0 = 0.$$

3. Equations in a Field

1. Solve the equation $a + x = b$ for $x \in \mathbb{R}$.
2. Solve the equation $ax = b$ for $a \neq 0$.

Solution:

(1) To solve

$$a + x = b$$

we add the additive inverse $-a$ to both sides. This gives

$$(-a) + (a + x) = (-a) + b.$$

By associativity, it follows that

$$((-a) + a) + x = (-a) + b.$$

Since $(-a) + a = 0$, we obtain

$$0 + x = (-a) + b,$$

hence

$$x = (-a) + b = b - a.$$

Thus the equation has the unique solution

$$x = b - a.$$

(2) Now let $a \neq 0$ and

$$ax = b.$$

We multiply both sides by the multiplicative inverse a^{-1}. It follows that

$$a^{-1}(ax) = a^{-1}b.$$

By associativity, we obtain

$$(a^{-1}a)x = a^{-1}b.$$

Since $a^{-1}a = 1$, it follows that

$$1 \cdot x = a^{-1}b.$$

By the neutrality of one, this gives

$$x = a^{-1}b.$$

Thus the equation has the unique solution

$$x = \frac{b}{a}.$$

4. Zero Product and Sign Rules

1. Show that $xy = 0 \Rightarrow x = 0$ or $y = 0$.

2. Show that $(-x)(-y) = xy$.

Solution:

(1) One direction is immediate: If $x = 0$ or $y = 0$, then, since $x \cdot 0 = 0$, it follows at once that

$$xy = 0.$$

For the converse direction, assume that

$$xy = 0$$

holds, and additionally that $x \neq 0$. Then x has a multiplicative inverse x^{-1}. Multiplying the equation $xy = 0$ by x^{-1} gives

$$x^{-1}(xy) = x^{-1} \cdot 0.$$

Using associativity and the already proved statement $x^{-1} \cdot 0 = 0$, we obtain

$$(x^{-1}x)y = 0.$$

Since $x^{-1}x = 1$, we get

$$1 \cdot y = 0,$$

hence

$$y = 0.$$

Thus we have shown that

$$xy = 0 \Rightarrow x = 0 \text{ or } y = 0.$$

(2) From the already proved relation

$$-x = (-1)x$$

and similarly

$$-y = (-1)y$$

it follows that

$$(-x)(-y) = ((-1)x)((-1)y).$$

By associativity and commutativity of multiplication, we may write this as

$$(-1)(-1)\,xy.$$

It remains to show that

$$(-1)(-1) = 1.$$

For this, we use the fact that $1 + (-1) = 0$. Then

$$0 = (-1) \cdot 0 = (-1)(1 + (-1)).$$

By the distributive law, it follows that

$$(-1) \cdot 1 + (-1) \cdot (-1) = 0.$$

Since $(-1) \cdot 1 = -1$, we obtain

$$-1 + (-1)(-1) = 0.$$

Thus $(-1)(-1)$ is an additive inverse of -1, and therefore

$$(-1)(-1) = 1.$$

Consequently,

$$(-x)(-y) = 1 \cdot xy = xy.$$

Remark 121. *The calculation rules proved in this exercise section will be used constantly later on, often without being justified again. Precisely for this reason, it is worthwhile to carefully trace their derivation from the axioms once.*

The following exercises are directly based on the results developed in this chapter concerning the order axioms, the absolute value, and the Archimedean axiom. They are intended to strengthen confidence in working with inequalities, estimates, and statements about growth. The solutions are deliberately somewhat detailed so that the respective proof ideas remain clearly visible.

1. Basic Properties of the Order

1. Show, using the definition of $x < y$, that either $x < y$, $x = y$, or $x > y$ holds.

2. Show that $x < y$ implies $x + a < y + a$ for all $a \in \mathbb{R}$.

3. Show that $x < y$ implies $-x > -y$.

Solution:

(1) By definition,

$$x < y \iff y - x > 0.$$

We apply the trichotomy axiom to the number $y - x$. Exactly one of the three statements

$$y - x > 0, \qquad y - x = 0, \qquad -(y - x) > 0$$

is true. In the first case, $x < y$ holds; in the second case, $x = y$; and in the third case,

$$-(y - x) = x - y > 0,$$

so $x > y$. Hence, for any two real numbers, exactly one of the three possibilities

$$x < y, \qquad x = y, \qquad x > y.$$

holds.

(2) From $x < y$ it follows by definition that

$$y - x > 0.$$

Now we consider the difference of the shifted numbers:

$$(y + a) - (x + a) = y - x.$$

Therefore,

$$(y + a) - (x + a) > 0$$

also holds, and by definition it follows that

$$x + a < y + a.$$

(3) From $x < y$ it again follows that

$$y - x > 0.$$

For the negated numbers, we have

$$(-x) - (-y) = y - x.$$

Thus,

$$(-x) - (-y) > 0.$$

By the definition of the order, this implies

$$-x > -y.$$

2. Monotonicity of multiplication

1. Show: From $x < y$ and $a > 0$ it follows that $ax < ay$.
2. Show: From $x < y$ and $a < 0$ it follows that $ax > ay$.
3. Show: From $0 \leq x < y$ and $0 \leq a < b$ it follows that $ax < by$.

Solution:

(1) From $x < y$ it follows that

$$y - x > 0.$$

Since, in addition, $a > 0$, the closure of positive numbers under multiplication gives

$$a(y - x) > 0.$$

By the distributive law,

$$a(y - x) = ay - ax.$$

Hence

$$ay - ax > 0,$$

and therefore, by definition,

$$ax < ay.$$

(2) From $a < 0$, changing signs gives

$$-a > 0.$$

Applying part (1) to the positive number $-a$, we obtain from $x < y$ that

$$(-a)x < (-a)y.$$

This is equivalent to

$$-ax < -ay.$$

Changing signs once more yields

$$ax > ay.$$

(3) We distinguish two cases.

Case 1: $a = 0$. Then

$$ax = 0.$$

Since, moreover, $b > 0$ and $y > 0$ follows from $0 \leq x < y$, part (1), applied with $0 < y$ and $b > 0$, also gives

$$0 < by.$$

Thus, in this case,

$$ax = 0 < by.$$

Case 2: $a > 0$. Then part (1), applied to $x < y$, gives

$$ax < ay.$$

Furthermore, applying part (1) again to $a < b$ and $y > 0$ gives

$$ay < by.$$

By the transitivity of $<$, we obtain

$$ax < by.$$

Since $ax < by$ holds in both cases, the assertion is proved.

3. Properties of the Absolute Value

1. Show that $|x| \geq 0$ and $|x| = 0 \Leftrightarrow x = 0$.

2. Show that $|xy| = |x|\,|y|$.

3. Show that $|x + y| \leq |x| + |y|$.

4. Show that $||x| - |y|| \leq |x - y|$.

Solution:

(1) By definition, either $|x| = x$ with $x \geq 0$, or $|x| = -x$ with $x < 0$. In both cases we have

$$|x| \geq 0.$$

Conversely, if $|x| = 0$, then the following follows:

- In the case $x \geq 0$, we have $|x| = x$, hence $x = 0$.
- In the case $x < 0$, we have $|x| = -x = 0$, hence again $x = 0$.

Thus

$$|x| = 0 \Leftrightarrow x = 0.$$

(2) We distinguish cases according to the signs of x and y. If x is nonnegative, then $|x| = x$; otherwise $|x| = -x$. The same applies to y. In all four possible sign cases, one obtains directly

$$|xy| = |x|\,|y|.$$

For example, in the case $x < 0$ and $y \geq 0$, we have

$$|xy| = -(xy) = (-x)y = |x|\,|y|.$$

The remaining cases are just as simple. This proves the multiplicativity of the absolute value.

(3) From the triangle inequality it follows that

$$|x| = |(x + y) - y| \leq |x + y| + |y|.$$

Subtracting $|y|$ from both sides gives

$$|x| - |y| \leq |x + y|.$$

Now interchanging the roles of x and y, or directly considering

$$|x + y| = |(x + y) - 0| \leq |x| + |y|,$$

we obtain the desired inequality

$$|x + y| \leq |x| + |y|.$$

(4) From the triangle inequality we first obtain

$$|x| = |(x - y) + y| \leq |x - y| + |y|.$$

It follows that

$$|x| - |y| \leq |x - y|.$$

Interchanging x and y gives

$$|y| - |x| \leq |y - x| = |x - y|.$$

Together, these two inequalities mean precisely that

$$||x| - |y|| \leq |x - y|.$$

4. Archimedean Axiom

1. Show that for every $x \in \mathbb{R}$ there exist $n_1, n_2 \in \mathbb{N}$ such that $n_1 > x$ and $-n_2 < x$.

2. Show that for every $x \in \mathbb{R}$ there exists exactly one $n \in \mathbb{Z}$ such that $n \leq x < n + 1$.

3. Show that for every $\varepsilon > 0$ there exists $n \in \mathbb{N}$ such that $\frac{1}{n} < \varepsilon$.

Solution:

(1) The Archimedean axiom states that for every real number x there exists a natural number n such that $n > x$. Applying this to x itself, we obtain an $n_1 \in \mathbb{N}$ with

$$n_1 > x.$$

Applying the Archimedean axiom also to $-x$, there exists an $n_2 \in \mathbb{N}$ with

$$n_2 > -x.$$

This is equivalent to

$$-n_2 < x.$$

Thus the two required natural numbers have been found.

(2) The existence of such an integer follows from the Archimedean axiom and from the fact that the integers divide the real line discretely into unit intervals. There is exactly one $n \in \mathbb{Z}$ such that

$$n \leq x < n + 1.$$

Uniqueness is seen as follows: If there were two integers n and m with

$$n \leq x < n+1 \qquad \text{and} \qquad m \leq x < m+1,$$

and if, for instance, $n < m$, then, since n and m are integers, we would have

$$n+1 \leq m.$$

Together with $m \leq x < n+1$, this would imply

$$n+1 \leq x < n+1,$$

a contradiction. Hence one must have $n = m$. The integer is therefore uniquely determined.

(3) Let $\varepsilon > 0$. Then also $\varepsilon^{-1} > 0$. By the Archimedean axiom, there exists an $n \in \mathbb{N}$ with

$$n > \varepsilon^{-1}.$$

Since $n > 0$ and the function $t \mapsto t^{-1}$ is strictly decreasing on the positive real numbers, it follows that

$$\frac{1}{n} < \varepsilon.$$

This proves the assertion.

5. Bernoulli and Growth

1. Prove Bernoulli's inequality: For $x \geq -1$ and $n \in \mathbb{N}_0$, one has

$$(1+x)^n \geq 1 + nx.$$

2. Prove: For $b > 1$, one has

$$\forall r \in \mathbb{R}\ \exists n \in \mathbb{N}: \ b^n > r.$$

3. Prove: For $0 < b < 1$, one has

$$\forall \varepsilon > 0\ \exists n \in \mathbb{N}: \ b^n < \varepsilon.$$

Solution:

(1) We prove the statement by complete induction on n. For $n = 0$, we have

$$(1 + x)^0 = 1 = 1 + 0 \cdot x.$$

Thus the base case is satisfied.

Assume that the assertion holds for a fixed $n \in \mathbb{N}_0$, that is,

$$(1 + x)^n \geq 1 + nx.$$

Since $x \geq -1$, we have

$$1 + x \geq 0.$$

Multiplying the induction hypothesis by $1 + x$ therefore preserves the inequality and gives

$$(1 + x)^{n+1} = (1 + x)^n(1 + x) \geq (1 + nx)(1 + x).$$

Expanding yields

$$(1 + nx)(1 + x) = 1 + (n + 1)x + nx^2.$$

Since $x^2 \geq 0$ and $n \in \mathbb{N}_0$, it follows that

$$nx^2 \geq 0.$$

Hence

$$(1+x)^{n+1} \geq 1 + (n+1)x.$$

This proves Bernoulli's inequality.

(2) Let $b > 1$. Then

$$b = 1 + x \qquad \text{with} \qquad x = b - 1 > 0.$$

By Bernoulli's inequality, for all $n \in \mathbb{N}_0$ we have

$$b^n = (1+x)^n \geq 1 + nx.$$

Now let $r \in \mathbb{R}$ be arbitrary. Since $x > 0$, the Archimedean axiom gives an $n \in \mathbb{N}$ such that

$$n > \frac{r-1}{x}.$$

Multiplication by x gives

$$nx > r - 1.$$

Therefore

$$1 + nx > r.$$

Together with $b^n \geq 1 + nx$, we obtain

$$b^n > r.$$

Thus the assertion is proved.

(3) Let $0 < b < 1$. Then

$$c := \frac{1}{b} > 1.$$

By part (2), for every $\varepsilon > 0$ there exists an $n \in \mathbb{N}$ such that

$$c^n > \frac{1}{\varepsilon}.$$

Since $c > 0$ and the monotonicity of reciprocals on the positive numbers has already been proved, it follows that

$$\frac{1}{c^n} < \varepsilon.$$

But

$$\frac{1}{c^n} = \left(\frac{1}{c}\right)^n = b^n.$$

Thus we obtain

$$b^n < \varepsilon.$$

This proves the statement.

Remark 122. *The exercises in this section form the technical foundation for later proofs involving limits and convergence. In particular, Bernoulli's inequality and the Archimedean axiom will be needed there again and again in precisely this form.*

Chapter 5

Sequences and Limits

The concepts of a *sequence* and a *limit* are among the fundamental ideas of analysis. Many of the quantities with which we work in mathematics cannot be captured exactly in a single finite step, but only approximated step by step with increasing accuracy. This is of central importance in analysis: instead of determining a value directly, one often considers a sequence of approximations that approaches a particular target value more and more closely. Such a target value is then called a *limit*.

This approach appears in many different contexts. In the computation of roots, in decimal expansions, in numerical methods, or in the study of functions, sequences are frequently used to describe mathematical quantities precisely. For example, the number $\sqrt{2}$ cannot be written as a finite decimal, but it can be approximated arbitrarily well by a sequence of rational numbers. The number π

can also be described in this way. The concept of a limit therefore provides the tool for formulating infinity and approximation in a mathematically precise manner.

A *sequence of real numbers* is, first of all, a mapping

$$a : \mathbb{N} \to \mathbb{R},$$

that is, an assignment that associates with each natural number $n \in \mathbb{N}$ a real number $a_n \in \mathbb{R}$. Instead of $a(n)$, one usually writes simply a_n. Such a sequence is denoted by

$$(a_n)_{n \in \mathbb{N}},$$

or also by

$$(a_0, a_1, a_2, \dots)$$

or simply by (a_n) if it is clear from the context which indices are meant.

Often a sequence does not begin at 0 or 1, but only from a specified starting index $k \in \mathbb{N}$. In that case one writes, for example,

$$(a_n)_{n \geq k}$$

or

$$(a_k, a_{k+1}, a_{k+2}, \dots).$$

In this case, a number a_n is given for every $n \geq k$, while the earlier indices play no role.

In what follows, we first consider a few simple examples in order to illustrate the concept of a sequence. After that, we introduce and make precise the concept of a limit.

5.1 Examples of Sequences

1. **Constant sequence:** For a fixed number $c \in \mathbb{R}$, one can consider the sequence

$$a_n = c \qquad (n \in \mathbb{N})$$

   All terms of this sequence are equal. Examples are

$$(1, 1, 1, 1, \dots) \qquad \text{or} \qquad (0, 0, 0, 0, \dots).$$

2. **Arithmetic sequence:** The sequence

$$a_n = a_0 + n \cdot d$$

   describes a linear progression with constant difference d. A simple example is

$$(0, 2, 4, 6, 8, \dots),$$

   that is, $a_n = 2n$.

3. **Geometric sequence:** Here each term of the sequence is obtained by multiplication with a fixed factor. An example is

$$\left(1, \frac{1}{2}, \frac{1}{4}, \frac{1}{8}, \dots\right),$$

   that is,

$$a_n = \left(\frac{1}{2}\right)^n.$$

4. **Sequence of reciprocals:** The sequence

$$a_n = \frac{1}{n+1}$$

gives the values

$$\left(1, \frac{1}{2}, \frac{1}{3}, \frac{1}{4}, \dots\right).$$

Later, it will be an important standard example of a convergent sequence.

5. **Alternating sequence:** The sequence

$$a_n = (-1)^n$$

alternately takes the values 1 and -1:

$$(1, -1, 1, -1, 1, -1, \dots).$$

This example shows that not every sequence must have a limit.

6. **Increasing sequence:** The sequence

$$a_n = n$$

is given by

$$(0, 1, 2, 3, 4, \dots)$$

or, depending on the starting index, also by

$$(1, 2, 3, 4, \dots).$$

It grows without bound and has no finite limit.

7. **Approximations of an irrational number:** The number $\sqrt{2}$ can be approximated more and more accurately by the sequence

$$1,\ 1.4,\ 1.41,\ 1.414,\ 1.4142, \ldots$$

Such sequences illustrate particularly clearly how a limit can be understood as the target of an ongoing approximation process.

These examples already make clear that sequences can exhibit very different kinds of behavior: some remain constant, some increase or decrease, others oscillate, and still others approach a fixed value. The task of analysis is now to describe this behavior precisely. For this purpose, we introduce the concept of a limit.

Definition 123 (Convergence of a sequence). *Let* $(a_n)_{n\in\mathbb{N}}$ *be a sequence of real numbers, and let* $a \in \mathbb{R}$. *The sequence is called* convergent to a *if*

$$\forall \varepsilon > 0\ \exists N \in \mathbb{N}\ \forall n \in \mathbb{N} :\ n \geq N \Rightarrow |a_n - a| < \varepsilon.$$

In this case one also writes

$$a_n \to a \quad (n \to \infty).$$

The idea behind this definition is the following: For every arbitrarily small $\varepsilon > 0$, *there is an index* N *such that all terms of the sequence from this index onward have distance less than* ε *from* a. *The index* N *may depend on* ε *and will, in general, become larger the smaller* ε *is chosen.*

Definition 124 (Limit, limes, null sequence). *If the sequence* $(a_n)_{n\in\mathbb{N}}$ *converges to* $a \in \mathbb{R}$*, then* a *is called the* limit *or* limes *of the sequence. One writes*

$$\lim_{n\to\infty} a_n = a \qquad \textit{or briefly} \qquad \lim a_n = a.$$

A sequence that converges to 0 *is called a* null sequence.

5.2 Geometric Interpretation

The definition of convergence can be interpreted very intuitively on the number line. Instead of working only with inequalities, one considers an arbitrarily small neighborhood around a given value $a \in \mathbb{R}$ into which all terms of the sequence should eventually fall.

Definition 125 (ε-neighborhood). *Let* $a \in \mathbb{R}$ *and* $\varepsilon > 0$. *Then*

$$U_\varepsilon(a) := \{x \in \mathbb{R} : |x - a| < \varepsilon\}$$

is called the ε-neighborhood *of* a. *It is precisely the open interval*

$$]a - \varepsilon,\ a + \varepsilon[$$

on the number line.

Using this terminology, the convergence of a sequence can be formulated as follows: A sequence $(a_n)_{n\in\mathbb{N}}$ converges to $a \in \mathbb{R}$ if and only if for every $\varepsilon > 0$ there exists an index $N \in \mathbb{N}$ such that for all $n \geq N$ one has:

$$a_n \in]a - \varepsilon,\ a + \varepsilon[\qquad \text{respectively} \qquad |a_n - a| < \varepsilon.$$

In other words: From some term onward, all terms of the sequence lie inside every arbitrarily small ε-neighborhood of a.

Intuitively, this means that the sequence approaches the value a so closely that, from some sufficiently large index onward, its terms can no longer be found outside a prescribed small neighborhood of a. It does not matter whether some early terms of the sequence are still far away; only the behavior of the sequence from a certain point onward is decisive.

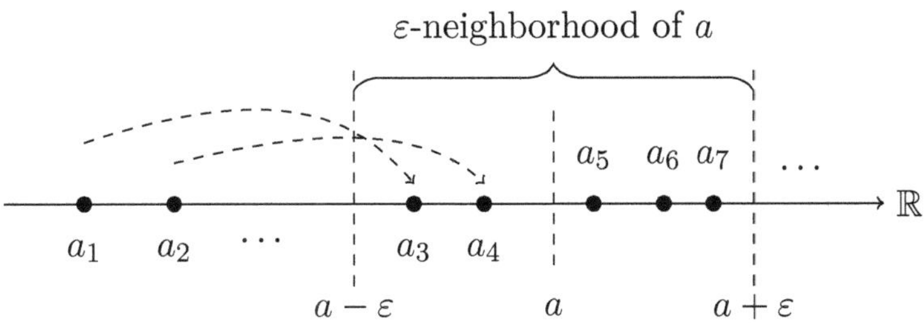

Figure 5.1: Geometric interpretation of convergence: From a certain index onward, all terms of the sequence lie inside every prescribed ε-neighborhood of the limit.

Definition 126 (Divergence)**.** *A sequence that does not have a limit is called* divergent.

5.3 Examples

In the following, we consider several simple but fundamental sequences and determine their limits. At the same time, the examples illustrate various typical types of behavior: constant behavior, convergence toward a fixed value, unbounded growth, and lack of convergence.

1. **Constant sequence**

   Let $c \in \mathbb{R}$ be fixed and

   $$a_n = c \qquad (n \in \mathbb{N}).$$

   Then, for every $\varepsilon > 0$ and every $n \in \mathbb{N}$, we already have

   $$|a_n - c| = |c - c| = 0 < \varepsilon.$$

   Hence

   $$\lim_{n\to\infty} a_n = c.$$

   Thus a constant sequence converges to its common value.

2. **Sequence of reciprocals**

   Consider the sequence

   $$a_n = \frac{1}{n+1} \qquad (n \in \mathbb{N}).$$

   We expect this sequence to converge to 0. To prove this, let $\varepsilon > 0$ be given. Then choose $N \in \mathbb{N}$ such that

   $$N > \frac{1}{\varepsilon} - 1.$$

Then, for all $n \geq N$, we have

$$0 < \frac{1}{n+1} \leq \frac{1}{N+1} < \varepsilon.$$

Therefore

$$\left| \frac{1}{n+1} - 0 \right| < \varepsilon \qquad \text{for all } n \geq N.$$

Hence

$$\lim_{n \to \infty} \frac{1}{n+1} = 0.$$

This sequence is a classical example of a null sequence.

3. **A sequence with limit 1**

Now consider

$$a_n = \frac{n}{n+1}.$$

It is useful to rewrite the expression:

$$\frac{n}{n+1} = 1 - \frac{1}{n+1}.$$

Since we already know that $\frac{1}{n+1} \to 0$, it is natural to expect that $a_n \to 1$. Indeed, for every $\varepsilon > 0$ we have:

$$\left| \frac{n}{n+1} - 1 \right| = \left| -\frac{1}{n+1} \right| = \frac{1}{n+1}.$$

Thus, if we again choose $N \in \mathbb{N}$ such that

$$N > \frac{1}{\varepsilon} - 1,$$

then it follows for all $n \geq N$ that

$$\left|\frac{n}{n+1} - 1\right| < \varepsilon.$$

Hence we obtain

$$\lim_{n\to\infty} \frac{n}{n+1} = 1.$$

4. **Geometric sequence**

   For a number q with $|q| < 1$, one considers the geometric sequence

   $$a_n = q^n.$$

   A particularly simple example is

   $$a_n = \left(\frac{1}{2}\right)^n = 1,\ \frac{1}{2},\ \frac{1}{4},\ \frac{1}{8}, \ldots$$

   Since the terms of this sequence become smaller and smaller and approach zero, we have

   $$\lim_{n\to\infty} \left(\frac{1}{2}\right)^n = 0.$$

   Thus geometric sequences with factor $|q| < 1$ are null sequences.

5. **A sequence without a limit**

   Now consider the sequence

   $$a_n = (-1)^n.$$

Its terms are

$$1, -1, 1, -1, 1, -1, \ldots$$

Thus the sequence jumps back and forth between two values. Therefore, it cannot approach any fixed limit. In fact, it is not convergent.

This can also be seen intuitively from the fact that there is no index from which onward the sequence remains in every arbitrarily small neighborhood of a single point. Consequently,

$$\lim_{n \to \infty} (-1)^n$$

is not defined.

The examples show that the concept of a limit describes the long-term behavior of a sequence. Some sequences approach a fixed value, others tend to 0, and still others have no limit at all.

5.4 Boundedness of Sequences

In addition to convergence, boundedness is an important property of a sequence. It describes whether the terms of the sequence remain within a fixed numerical range, or whether they extend without bound upward or downward. Intuitively, boundedness means that the sequence does not attain arbitrarily large absolute values.

Definition 127 (Boundedness of sequences). *Let* $(a_n)_{n\in\mathbb{N}}$ *be a sequence of real numbers.*

The sequence is called bounded above *if there exists a real number* $K \in \mathbb{R}$ *such that*

$$a_n \leq K \qquad \text{for all } n \in \mathbb{N}.$$

It is called bounded below *if there exists a real number* $L \in \mathbb{R}$ *such that*

$$a_n \geq L \qquad \text{for all } n \in \mathbb{N}.$$

The sequence is called bounded *if there exists a number* $M > 0$ *such that*

$$|a_n| \leq M \qquad \text{for all } n \in \mathbb{N}.$$

Thus a sequence is bounded precisely when its terms lie between two fixed bounds. The condition $|a_n| \leq M$ combines these two bounds into a single statement.

Theorem 128. *Every convergent sequence of real numbers is bounded.*

Proof. Let $(a_n)_{n\in\mathbb{N}}$ be a sequence such that

$$\lim_{n\to\infty} a_n = a.$$

By the definition of convergence, there exists an $N \in \mathbb{N}$ such that, for all $n \geq N$,

$$|a_n - a| < 1.$$

Using the triangle inequality, it follows that

$$|a_n| = |(a_n - a) + a| \leq |a_n - a| + |a| < |a| + 1 \qquad \text{for all } n \geq N.$$

The first finitely many terms $a_0, a_1, \ldots, a_{N-1}$ are also all finite. Therefore the number

$$M := \max\{|a_0|, |a_1|, \ldots, |a_{N-1}|, |a| + 1\}$$

is well defined and satisfies

$$|a_n| \leq M \qquad \text{for all } n \in \mathbb{N}.$$

Hence the sequence is bounded. □

Remark 129. *The converse does not hold in general: not every bounded sequence is convergent. A simple counterexample is the sequence*

$$a_n = (-1)^n,$$

since it is bounded, because $|a_n| = 1$ *always holds, but it has no limit.*

Example 130 (Bounded convergent sequence)**.** *Consider the sequence*

$$a_n = \frac{1}{n+1} \qquad (n \in \mathbb{N}).$$

Then $0 < a_n \leq 1$ *holds for all* $n \in \mathbb{N}$*. Thus the sequence is bounded. Moreover, it converges to* 0*, because for large* n *the terms of the sequence become arbitrarily small:*

$$\lim_{n \to \infty} \frac{1}{n+1} = 0.$$

This is a typical example of a sequence that is both bounded and convergent.

Example 131 (Unbounded sequence). *Consider the sequence*

$$a_n = n \qquad (n \in \mathbb{N}).$$

Its terms grow upward without bound. For every real number K, there exists an $n \in \mathbb{N}$ such that $a_n = n > K$. Hence the sequence is not bounded. In particular, it has no finite limit.

5.5 Uniqueness of the Limit

An important point is that, if a sequence converges, it cannot have several different limits. The limit is therefore uniquely determined. This is by no means self-evident, but rather a central feature of the concept of convergence.

Theorem 132 (Uniqueness of the limit). *Let $(a_n)_{n\in\mathbb{N}}$ be a sequence of real numbers. If this sequence converges both to $a \in \mathbb{R}$ and to $b \in \mathbb{R}$, then*

$$a = b.$$

Proof. Assume that $a \neq b$. Then, in particular,

$$|a - b| > 0.$$

We set

$$\varepsilon := \frac{|a-b|}{2}.$$

Since the sequence (a_n) converges to a by assumption, there exists an $N_1 \in \mathbb{N}$ such that

$$|a_n - a| < \varepsilon \qquad \text{for all } n \geq N_1.$$

Similarly, the convergence to b implies that there exists an $N_2 \in \mathbb{N}$ such that

$$|a_n - b| < \varepsilon \qquad \text{for all } n \geq N_2.$$

For every

$$n \geq \max\{N_1, N_2\}$$

both inequalities therefore hold simultaneously. By the triangle inequality, we then obtain

$$|a - b| \leq |a - a_n| + |a_n - b| < \varepsilon + \varepsilon = 2\varepsilon = |a - b|.$$

Thus it follows that

$$|a - b| < |a - b|,$$

a contradiction. Hence our assumption $a \neq b$ cannot be correct. Consequently,

$$a = b$$

must hold. □

Remark 133. *The uniqueness of the limit shows that it is meaningful to speak of* the *limit of a convergent sequence.*

5.6 Limit Laws

Once the limits of simple sequences have been determined, they can often be used to compute the limits of more complicated sequences. The rules for sums and products are especially important in this context. They show that the concept of a limit is compatible with the basic algebraic operations.

5.6.1 Sums of Convergent Sequences

Theorem 134 (Limit of a sum). *Let $(a_n)_{n\in\mathbb{N}}$ and $(b_n)_{n\in\mathbb{N}}$ be two sequences of real numbers such that*

$$\lim_{n\to\infty} a_n = a \qquad \text{and} \qquad \lim_{n\to\infty} b_n = b.$$

Then the sequence $(a_n + b_n)_{n\in\mathbb{N}}$ also converges, and

$$\lim_{n\to\infty} (a_n + b_n) = a + b.$$

Proof. Let $\varepsilon > 0$ be arbitrary. Since $a_n \to a$ and $b_n \to b$, there exist indices $N_1, N_2 \in \mathbb{N}$ such that

$$|a_n - a| < \frac{\varepsilon}{2} \qquad \text{for all } n \geq N_1$$

and

$$|b_n - b| < \frac{\varepsilon}{2} \qquad \text{for all } n \geq N_2.$$

For

$$N := \max\{N_1, N_2\}$$

it then follows for all $n \geq N$ that

$$|(a_n + b_n) - (a + b)| = |(a_n - a) + (b_n - b)|.$$

By the triangle inequality, we obtain

$$|(a_n + b_n) - (a + b)| \leq |a_n - a| + |b_n - b| < \frac{\varepsilon}{2} + \frac{\varepsilon}{2} = \varepsilon.$$

Thus $(a_n + b_n)$ converges to $a + b$. □

5.6.2 Products of Convergent Sequences

Theorem 135 (Limit of a product). *Let $(a_n)_{n\in\mathbb{N}}$ and $(b_n)_{n\in\mathbb{N}}$ be two sequences of real numbers such that*

$$\lim_{n\to\infty} a_n = a \qquad \textit{and} \qquad \lim_{n\to\infty} b_n = b.$$

Then the sequence $(a_n b_n)_{n\in\mathbb{N}}$ also converges, and

$$\lim_{n\to\infty} (a_n b_n) = ab.$$

Proof. Since (a_n) converges, the sequence is bounded by the theorem already proved. Hence there exists a constant $M > 0$ such that

$$|a_n| \leq M \qquad \text{for all } n \in \mathbb{N}.$$

Moreover, (b_n) converges to b.

For each n, we now consider the difference

$$a_n b_n - ab = a_n(b_n - b) + b(a_n - a).$$

By the triangle inequality, it follows that

$$|a_n b_n - ab| \leq |a_n|\,|b_n - b| + |b|\,|a_n - a|.$$

Let $\varepsilon > 0$ be given. Since $b_n \to b$, there exists an $N_1 \in \mathbb{N}$ such that

$$|b_n - b| < \frac{\varepsilon}{2M} \qquad \text{for all } n \geq N_1.$$

Since $a_n \to a$, there exists an $N_2 \in \mathbb{N}$ such that

$$|a_n - a| < \frac{\varepsilon}{2(|b|+1)} \qquad \text{for all } n \geq N_2.$$

For

$$N := \max\{N_1, N_2\}$$

we then have, for all $n \geq N$:

$$|a_n b_n - ab| \leq M \cdot \frac{\varepsilon}{2M} + |b| \cdot \frac{\varepsilon}{2(|b|+1)} < \frac{\varepsilon}{2} + \frac{\varepsilon}{2} = \varepsilon.$$

This shows that

$$\lim_{n\to\infty} (a_n b_n) = ab.$$

□

Remark 136. *With the help of these two rules, the limits of many sequences can be computed step by step. Together with the later rules for quotients and powers, they form a very useful collection of computational tools for analysis.*

Corollary 137 (Linear combinations of convergent sequences). *Let* $(a_n)_{n\in\mathbb{N}}$ *and* $(b_n)_{n\in\mathbb{N}}$ *be two convergent sequences of real numbers, and let* $\lambda, \mu \in \mathbb{R}$. *Then the sequence*

$$(\lambda a_n + \mu b_n)_{n\in\mathbb{N}},$$

also converges, and

$$\lim_{n\to\infty}(\lambda a_n + \mu b_n) = \lambda \lim_{n\to\infty} a_n + \mu \lim_{n\to\infty} b_n.$$

Proof. Since (a_n) and (b_n) converge, the previously proved limit laws also imply that the sequences (λa_n) and (μb_n) converge. Moreover,

$$\lim_{n\to\infty}(\lambda a_n) = \lambda \lim_{n\to\infty} a_n$$

and

$$\lim_{n\to\infty}(\mu b_n) = \mu \lim_{n\to\infty} b_n.$$

Finally, by the sum rule,

$$\lim_{n\to\infty}(\lambda a_n + \mu b_n) = \lambda \lim_{n\to\infty} a_n + \mu \lim_{n\to\infty} b_n.$$

□

5.6.3 Quotient of convergent sequences

Quotients can also be handled well, provided that the denominator does not tend to zero. The central idea is this: if a sequence (b_n) converges to a number b different from 0, then the terms b_n themselves are also nonzero for sufficiently large n, and their reciprocals approach the reciprocal of b. Thus a quotient can be written as a product, and the product rule already proved can be applied.

Theorem 138 (Limit of a quotient). *Let $(a_n)_{n\in\mathbb{N}}$ and $(b_n)_{n\in\mathbb{N}}$ be two sequences of real numbers with*

$$\lim_{n\to\infty} a_n = a \qquad \textit{and} \qquad \lim_{n\to\infty} b_n = b,$$

where $b \neq 0$. Then $b_n \neq 0$ for all sufficiently large n, the sequence

$$\left(\frac{a_n}{b_n}\right)_{n\in\mathbb{N}}$$

is convergent, and

$$\lim_{n\to\infty} \frac{a_n}{b_n} = \frac{a}{b}.$$

Proof. We first show that the sequence of reciprocals $(1/b_n)$ converges to $1/b$.

Since $b_n \to b$ and $b \neq 0$, there exists an $N_0 \in \mathbb{N}$ such that

$$|b_n - b| < \frac{|b|}{2} \qquad \text{for all } n \geq N_0.$$

For such n, we then have in particular

$$|b_n| \geq |b| - |b_n - b| > |b| - \frac{|b|}{2} = \frac{|b|}{2}.$$

Thus $b_n \neq 0$ for all $n \geq N_0$.

Moreover, for all $n \geq N_0$ we obtain

$$\left|\frac{1}{b_n} - \frac{1}{b}\right| = \frac{|b - b_n|}{|b_n|\,|b|} \leq \frac{2}{|b|^2}\,|b_n - b|.$$

Since $|b_n - b| \to 0$, it follows that

$$\frac{1}{b_n} \to \frac{1}{b}.$$

Now, for $n \geq N_0$, we write

$$\frac{a_n}{b_n} = a_n \cdot \frac{1}{b_n}.$$

Since $a_n \to a$ and $\frac{1}{b_n} \to \frac{1}{b}$, the product rule for convergent sequences gives

$$\lim_{n\to\infty} \frac{a_n}{b_n} = \left(\lim_{n\to\infty} a_n\right) \cdot \left(\lim_{n\to\infty} \frac{1}{b_n}\right) = a \cdot \frac{1}{b} = \frac{a}{b}.$$

This proves the assertion. □

Remark 139. *The assumption $b \neq 0$ is essential. If the limit of the denominator is equal to 0, then a quotient need not converge; it may, for example, become unbounded or oscillate.*

5.6.4 Monotonicity Behavior of Limits

Limits preserve the order of the real numbers: if one sequence always remains below another, then its limit also remains below the limit of the other sequence. This principle is very useful because many estimates can be transferred directly to limits in this way.

Theorem 140 (Order preservation of limits). *Let $(a_n)_{n\in\mathbb{N}}$ and $(b_n)_{n\in\mathbb{N}}$ be two convergent sequences of real numbers such that*

$$a_n \leq b_n \qquad \text{for all } n \in \mathbb{N}.$$

Then also

$$\lim_{n\to\infty} a_n \leq \lim_{n\to\infty} b_n.$$

Proof. Let

$$\lim_{n\to\infty} a_n = a \qquad \text{and} \qquad \lim_{n\to\infty} b_n = b.$$

We show that necessarily $a \leq b$.

Suppose, to the contrary, that $a > b$. Then

$$a - b > 0.$$

Set

$$\varepsilon := \frac{a-b}{2} > 0.$$

Since $a_n \to a$, there exists an $N_1 \in \mathbb{N}$ such that

$$|a_n - a| < \varepsilon \qquad \text{for all } n \geq N_1,$$

and since $b_n \to b$, there exists an $N_2 \in \mathbb{N}$ such that

$$|b_n - b| < \varepsilon \qquad \text{for all } n \geq N_2.$$

For

$$n \geq \max\{N_1, N_2\}$$

it follows that

$$a_n > a - \varepsilon \qquad \text{and} \qquad b_n < b + \varepsilon.$$

By the choice of ε, we have

$$a - \varepsilon = a - \frac{a-b}{2} = \frac{a+b}{2} \quad \text{and} \quad b + \varepsilon = b + \frac{a-b}{2} = \frac{a+b}{2}.$$

Hence, for all sufficiently large n,

$$a_n > \frac{a+b}{2} > b_n,$$

which contradicts the assumption that $a_n \leq b_n$ for all n.

Thus the assumption $a > b$ cannot be true. Consequently,

$$a \leq b.$$

□

Corollary 141 (Enclosure by bounds). *Let $x, y \in \mathbb{R}$ and let $(a_n)_{n \in \mathbb{N}}$ be a convergent sequence of real numbers such that*

$$x \leq a_n \leq y \qquad \textit{for all } n \in \mathbb{N}.$$

Then

$$x \leq \lim_{n \to \infty} a_n \leq y.$$

Proof. The constant numbers x and y can be regarded as constant sequences

$$x_n := x \qquad \text{and} \qquad y_n := y$$

Then, for all $n \in \mathbb{N}$, we have

$$x_n \leq a_n \leq y_n.$$

Since constant sequences converge to their constant value, we have

$$\lim_{n\to\infty} x_n = x \qquad \text{and} \qquad \lim_{n\to\infty} y_n = y.$$

By the preceding theorem, it follows that

$$x \leq \lim_{n\to\infty} a_n \leq y.$$

□

5.7 Divergence to $\pm\infty$

Not every sequence has a finite limit. Some sequences instead grow without an upper bound or decrease without a lower bound. To describe this behavior precisely, one considers a special form of divergence in which the terms of the sequence do not tend to a real number, but to $+\infty$ or $-\infty$. This is also called *proper divergence* or *improper convergence.*

Definition 142 (Divergence to $+\infty$ and $-\infty$). *A sequence $(a_n)_{n\in\mathbb{N}}$ of real numbers is said to be* properly divergent to $+\infty$ *if for every real number K there exists an index $N \in \mathbb{N}$ such that*

$$a_n > K \qquad \textit{for all } n \geq N.$$

In this case one writes

$$\lim_{n\to\infty} a_n = +\infty.$$

The sequence $(a_n)_{n\in\mathbb{N}}$ *is said to be* properly divergent to $-\infty$ *if the sequence* $(-a_n)_{n\in\mathbb{N}}$ *is properly divergent to* $+\infty$*. In that case one writes*

$$\lim_{n\to\infty} a_n = -\infty.$$

Intuitively, $\lim_{n\to\infty} a_n = +\infty$ means that the terms of the sequence eventually exceed every arbitrarily large bound. In the case $\lim_{n\to\infty} a_n = -\infty$, the terms behave correspondingly in the downward direction.

Example 143. *The sequence*

$$a_n = n$$

diverges to $+\infty$*, because for every real number* K *one can choose an* $N \in \mathbb{N}$ *such that* $N > K$*. Then, for all* $n \geq N$*, one also has* $a_n = n > K$*.*

Example 144. *The sequence*

$$a_n = n^2$$

also diverges to $+\infty$*. The same is true for many other increasing sequences, for example*

$$a_n = 2^n.$$

Example 145. *The sequence*

$$a_n = -n$$

diverges to $-\infty$*, because* $-a_n = n$ *diverges to* $+\infty$*.*

Remark 146. *Every sequence that diverges to* $+\infty$ *or to* $-\infty$ *is necessarily unbounded. Indeed, if* (a_n) *were bounded above, for instance, then there would be a number* K *such that* $a_n \leq K$ *for all* n*, contradicting the definition of* $\lim_{n\to\infty} a_n = +\infty$*.*

The converse, however, is not true: not every unbounded sequence diverges properly to $+\infty$ or to $-\infty$. A simple counterexample is

$$a_n = (-1)^n n.$$

This sequence is unbounded, since the absolute values of its terms become arbitrarily large. However, it diverges neither to $+\infty$ nor to $-\infty$, because it alternately takes large positive and large negative values.

Remark 147. *The symbols* $+\infty$ *and* $-\infty$ *are not real numbers. Indeed, there is no real number that is greater than all real numbers, and likewise no real number that is smaller than all real numbers. For every real number* x*, for instance, one has* $x + 1 > x$*, so no real number can play the role of* $+\infty$*. Similarly,* $x - 1 < x$ *shows that there can be no real* $-\infty$*.*

For many purposes, however, it is useful to include these two symbols and to consider the so-called *extended real number system*:

$$\hat{\mathbb{R}} := \mathbb{R} \cup \{-\infty, +\infty\}.$$

5.8 Reciprocals and Limits

The study of reciprocals reveals a close connection between null sequences and sequences that diverge to $+\infty$ or to $-\infty$. Intuitively, if the absolute values of the terms of a sequence become very large, then their reciprocals become very small. Conversely, the reciprocals of a null sequence that does not change sign become unboundedly large in the positive or negative sense.

Theorem 148 (Reciprocals of properly divergent sequences). *Let $(a_n)_{n\in\mathbb{N}}$ be a sequence of real numbers that properly diverges to $+\infty$ or to $-\infty$. Then there exists an $n_0 \in \mathbb{N}$ such that*

$$a_n \neq 0 \qquad \text{for all } n > n_0,$$

and the sequence of reciprocals converges to 0*, that is,*

$$\lim_{n\to\infty} \frac{1}{a_n} = 0.$$

Proof. We distinguish between the two cases.

Case 1: $\lim_{n\to\infty} a_n = +\infty$.

By definition, for every $K \in \mathbb{R}$ there exists an $N \in \mathbb{N}$ such that

$$a_n > K \qquad \text{for all } n \geq N.$$

Now let $\varepsilon > 0$ be given. We choose

$$K := \frac{1}{\varepsilon}.$$

Then there exists an $N \in \mathbb{N}$ such that

$$a_n > \frac{1}{\varepsilon} \qquad \text{for all } n \geq N.$$

In particular, $a_n \neq 0$ for all $n \geq N$, and moreover

$$\left|\frac{1}{a_n} - 0\right| = \frac{1}{|a_n|} < \varepsilon \qquad \text{for all } n \geq N.$$

Thus it follows that

$$\lim_{n\to\infty} \frac{1}{a_n} = 0.$$

Case 2: $\lim_{n\to\infty} a_n = -\infty$.

Then $(-a_n)$ diverges properly to $+\infty$. By the first case, we therefore have

$$\lim_{n\to\infty} \frac{1}{-a_n} = 0.$$

Since

$$\frac{1}{-a_n} = -\frac{1}{a_n},$$

it follows immediately that

$$\lim_{n\to\infty} \frac{1}{a_n} = 0.$$

Here, too, we have in particular $a_n \neq 0$ for all sufficiently large n. □

Theorem 149 (Reciprocals of null sequences with fixed sign). *Let $(a_n)_{n\in\mathbb{N}}$ be a null sequence.*

- *If $a_n > 0$ for all $n \in \mathbb{N}$, then the sequence $\left(\frac{1}{a_n}\right)_{n\in\mathbb{N}}$ diverges properly to $+\infty$.*
- *If $a_n < 0$ for all $n \in \mathbb{N}$, then the sequence $\left(\frac{1}{a_n}\right)_{n\in\mathbb{N}}$ diverges properly to $-\infty$.*

Proof. We again treat the two cases separately.

Case 1: $a_n > 0$ for all n.

Let $K \in \mathbb{R}$ be arbitrary. Since (a_n) converges to 0, for

$$\varepsilon := \frac{1}{K+1}$$

there exists an $N \in \mathbb{N}$ such that

$$|a_n| < \varepsilon \qquad \text{for all } n \geq N.$$

Since $a_n > 0$, it follows for all $n \geq N$ that

$$0 < a_n < \varepsilon.$$

Hence

$$\frac{1}{a_n} > \frac{1}{\varepsilon} = K + 1 > K \qquad \text{for all } n \geq N.$$

Thus $\left(\frac{1}{a_n}\right)$ diverges properly to $+\infty$.

Case 2: $a_n < 0$ for all n.

Then $(-a_n)$ is a positive null sequence. By the first case, we therefore have

$$\lim_{n\to\infty} \frac{1}{-a_n} = +\infty.$$

Since

$$\frac{1}{-a_n} = -\frac{1}{a_n}$$

it follows that

$$\lim_{n\to\infty} \frac{1}{a_n} = -\infty.$$

□

Example 150. *The sequence*

$$a_n = \frac{1}{n+1}$$

is a positive null sequence. Hence

$$\frac{1}{a_n} = n + 1,$$

and this sequence diverges properly to $+\infty$.

Conversely,

$$b_n = n + 1$$

is a sequence with $b_n \to +\infty$*, and its reciprocals form the null sequence*

$$\frac{1}{b_n} = \frac{1}{n+1} \to 0.$$

5.9 Exercises and Solutions

The following exercises are intended to consolidate the basic concepts and calculation rules for sequences. Some exercises check simple limit computations, while others train the use of divergence, boundedness, and estimates. The solutions are deliberately written in detail so that the underlying ideas become visible.

1. **Determine the limit of the constant sequence.**

   Let the sequence be given by

   $$a_n = 7 \qquad (n \in \mathbb{N}).$$

   Determine $\lim_{n\to\infty} a_n$.

   Solution. Since all terms of the sequence are equal to 7, the sequence does not change as n increases. For every $\varepsilon > 0$, we even have

   $$|a_n - 7| = |7 - 7| = 0 < \varepsilon \qquad \text{for all } n \in \mathbb{N}.$$

   Therefore, the sequence converges to 7. Thus

   $$\lim_{n\to\infty} a_n = 7.$$

2. **Determine the limit of a simple rational sequence.**

   Let
   $$b_n = \frac{n}{n+1}.$$
   Determine the limit of (b_n).

   Solution. First, we rewrite the expression:
   $$\frac{n}{n+1} = 1 - \frac{1}{n+1}.$$
   From the previous example we know that
   $$\frac{1}{n+1} \to 0.$$
   Therefore, it is natural to expect that b_n converges to 1. Formally, we see that
   $$\left|\frac{n}{n+1} - 1\right| = \left|-\frac{1}{n+1}\right| = \frac{1}{n+1}.$$
   This expression becomes arbitrarily small as n becomes large. Therefore,
   $$\lim_{n\to\infty} \frac{n}{n+1} = 1.$$

3. **Examine a geometric sequence.**

   Consider the sequence
   $$c_n = \left(\frac{2}{3}\right)^n.$$
   Determine its limit.

Solution. Since the factor $\frac{2}{3}$ has absolute value less than 1, the terms of the sequence become smaller and smaller. In particular,

$$0 < \left(\frac{2}{3}\right)^n \leq \left(\frac{2}{3}\right)^n$$

and the values approach 0. More precisely, this is a geometric sequence with ratio $q = \frac{2}{3}$ and $|q| < 1$. Such sequences are null sequences. Therefore,

$$\lim_{n\to\infty} \left(\frac{2}{3}\right)^n = 0.$$

4. **Show that a sequence is bounded.**

Examine the sequence

$$d_n = (-1)^n$$

with respect to boundedness.

Solution. The sequence alternates between 1 and -1. Hence, for every $n \in \mathbb{N}$,

$$|d_n| = |(-1)^n| = 1.$$

Therefore,

$$|d_n| \leq 1 \qquad \text{for all } n \in \mathbb{N}.$$

Thus the sequence is bounded, for example by the bound $M = 1$.

At the same time, this shows that boundedness alone does not guarantee convergence, since the sequence has no limit.

5. **Show divergence to $+\infty$.**

Consider the sequence

$$e_n = n^2.$$

Show that $\lim_{n\to\infty} e_n = +\infty$.

Solution. Let $K \in \mathbb{R}$ be arbitrary. We must find an $N \in \mathbb{N}$ such that, for all $n \geq N$,

$$n^2 > K.$$

Since the squares of the natural numbers become arbitrarily large, it suffices to choose N such that

$$N > \sqrt{K}$$

if $K \geq 0$. If $K < 0$, then the inequality holds automatically for all $n \in \mathbb{N}$, since always $n^2 \geq 0 > K$.

Thus e_n exceeds every prescribed bound. Hence,

$$\lim_{n\to\infty} n^2 = +\infty.$$

6. **Examine a quotient with known denominator limit.**

Let

$$f_n = \frac{3n+1}{n+2}.$$

Determine the limit of (f_n).

Solution. We divide numerator and denominator by n:

$$\frac{3n+1}{n+2} = \frac{3+\frac{1}{n}}{1+\frac{2}{n}}.$$

Now

$$\frac{1}{n} \to 0 \qquad \text{and} \qquad \frac{2}{n} \to 0.$$

Hence the numerator converges to 3 and the denominator to 1. By the quotient rule, it follows that

$$\lim_{n\to\infty} \frac{3n+1}{n+2} = \frac{3}{1} = 3.$$

7. **Use an estimate to determine a limit.**

   Let the sequence be given by

$$g_n = \frac{n}{n^2+1}.$$

   Show that (g_n) is a null sequence.

Solution. For $n \in \mathbb{N}$, we have

$$0 < \frac{n}{n^2+1} \leq \frac{n}{n^2} = \frac{1}{n}.$$

Since

$$\frac{1}{n} \to 0,$$

the squeeze theorem also implies that

$$\frac{n}{n^2+1} \to 0.$$

Thus the sequence is a null sequence.

8. **Examine the reciprocal of a null sequence with fixed sign.**

   Consider the sequence
   $$h_n = \frac{1}{n+1}.$$
   Determine the limit of the sequence
   $$\left(\frac{1}{h_n}\right)_{n\in\mathbb{N}}.$$

   Solution. First, we have
   $$h_n = \frac{1}{n+1} > 0 \qquad \text{for all } n \in \mathbb{N},$$
   and, moreover, (h_n) is a null sequence, since
   $$\frac{1}{n+1} \to 0.$$
   By the theorem on reciprocals of positive null sequences, the reciprocal sequence diverges properly to $+\infty$. Since
   $$\frac{1}{h_n} = n+1,$$
   it follows directly that
   $$\lim_{n\to\infty} \frac{1}{h_n} = +\infty.$$

The exercises show how the fundamental tools from this chapter work together: direct limit computation, transformations, estimates, boundedness, and divergence.

Chapter 6

The Completeness Axiom

After the chapters on the natural numbers, the field axioms, the order axioms, as well as sequences and limits, a central guiding question of analysis arises: Why is the number system of the real numbers so well suited for treating limiting processes, approximations, and questions of existence?

A first indication is already provided by the number $\sqrt{2}$. It can be approximated arbitrarily well by rational numbers, but it is not itself a rational number. This shows that the number system $\mathbb{Q}$, although it has many algebraic and order-theoretic structures, is not "closed" in every relevant sense. There are gaps that cannot be filled by rational numbers. The real numbers $\mathbb{R}$ close precisely these gaps.

This is also reflected in the extremal property: Consider, for instance, the set

$$A := \{ q \in \mathbb{Q} \mid q^2 < 2 \},$$

It is bounded above in $\mathbb{Q}$, but it has no supremum there. In $\mathbb{R}$, by contrast, a least upper bound does exist, namely $\sqrt{2}$. Thus the existence of suprema and infima is not a minor technical detail, but an expression of a fundamental completeness property of the real numbers.

It is precisely this completeness that makes $\mathbb{R}$ the central number system of analysis. It guarantees that limiting processes do not lead into a void, but actually arrive at a real limit. A particularly natural way to approach this idea is through *Cauchy sequences.* Here, convergence behavior is not described in terms of an already known limit, but only through the mutual approximation of the terms of the sequence. This is especially important from a didactic point of view, because it allows one to formulate a property of convergent sequences without already presupposing the limit.

6.1 Cauchy Sequences

Definition 151. *A sequence* $(a_n)_{n\in\mathbb{N}}$ *of real numbers is called a* Cauchy sequence *or* fundamental sequence *if for every* $\varepsilon > 0$ *there exists a natural number* $N \in \mathbb{N}$ *such that for all* $n, m \geq N$ *one has*

$$|a_n - a_m| < \varepsilon.$$

The intuition behind this definition is simple: from some sufficiently large index onward, the terms of the sequence lie arbitrarily close to one another. Thus it is no longer the size of the individual terms themselves that matters, but only the fact that their mutual distances become arbitrarily small. A Cauchy sequence "contracts"; it behaves as if it were moving toward a fixed point.

The connection with limits is fundamental:

Theorem 152. *Every convergent sequence of real numbers is a Cauchy sequence.*

Proof. Let (a_n) be a sequence such that

$$\lim_{n\to\infty} a_n = a.$$

For an arbitrary $\varepsilon > 0$, there exists an $N \in \mathbb{N}$ such that

$$|a_n - a| < \frac{\varepsilon}{2} \qquad \text{for all } n \geq N.$$

Hence, for arbitrary $n, m \geq N$, we obtain

$$|a_n - a_m| = |(a_n - a) - (a_m - a)| \leq |a_n - a| + |a_m - a| < \frac{\varepsilon}{2} + \frac{\varepsilon}{2} = \varepsilon$$

Thus (a_n) is a Cauchy sequence. □

This theorem shows that every convergent sequence is necessarily Cauchy. The converse, however, does not hold in every number system. In $\mathbb{Q}$ in particular, there are Cauchy sequences of rational numbers that do not have a rational limit. A classical example is a rational approximation of $\sqrt{2}$: the terms of the sequence move closer and closer together, but their limit no longer lies in $\mathbb{Q}$. This makes clear that the rational numbers are incomplete for the purposes of analysis.

This leads to the central completeness principle of the real numbers:

Axiom 153 (Completeness axiom)**.** *Every Cauchy sequence of real numbers converges in* $\mathbb{R}$*.*

This axiom characterizes $\mathbb{R}$ as a number system in which no limits are "missing", provided the sequence itself already becomes sufficiently close together. It is one of the most important properties of the real numbers and forms the basis for numerous further results in analysis.

6.1.1 The Irrationality of $\sqrt{2}$

A first and especially well-known example illustrating the significance of completeness is the number $\sqrt{2}$. Although it can be approximated arbitrarily well by rational numbers, it is not itself a rational number. The following classical proof shows this in an elementary way.

Theorem 154. *The number $\sqrt{2}$ is irrational.*

Proof. Assume that $\sqrt{2} \in \mathbb{Q}$. Then there exist relatively prime integers $n, m \in \mathbb{Z}$ with $m \neq 0$ such that

$$\sqrt{2} = \frac{n}{m}.$$

Squaring both sides gives

$$2 = \frac{n^2}{m^2}, \qquad \text{hence} \qquad n^2 = 2m^2.$$

Thus n^2 is even, and therefore n is even as well. Hence there exists a $k \in \mathbb{Z}$ such that $n = 2k$. Substitution yields

$$(2k)^2 = 2m^2 \qquad \Rightarrow \qquad 4k^2 = 2m^2 \qquad \Rightarrow \qquad m^2 = 2k^2.$$

Thus m is even as well. This contradicts the assumption that n and m are relatively prime. Consequently, $\sqrt{2}$ is irrational. □

This example already shows that the rational numbers cannot answer all natural questions of limits and existence in analysis. Although $\mathbb{Q}$ is an ordered field, it is not complete. This is precisely the decisive difference from the real numbers.

6.1.2 Why the completeness axiom is new

The completeness axiom cannot be derived from the field and order axioms considered so far. A simple counterexample is the number system $\mathbb{Q}$ itself: It satisfies the field axioms and the order axioms, but it is not complete. Indeed, there are Cauchy sequences of rational numbers that have no limit in $\mathbb{Q}$, for example rational approximations to $\sqrt{2}$.

This makes clear that completeness is not merely a reformulation of the previous axioms, but an additional fundamental structural idea. It turns the real numbers into a number system in which limiting processes can genuinely be carried through to completion.

6.1.3 Existence of square roots of positive real numbers

A first profound consequence of completeness is that positive real numbers have square roots.

Theorem 155. *For every number $a > 0$, there exists an $x \in \mathbb{R}$ such that*

$$x^2 = a.$$

Proof. We consider the set

$$A := \{\, t \in \mathbb{R} \mid t \geq 0 \text{ and } t^2 < a \,\}.$$

This set is nonempty, since $0 \in A$. Moreover, it is bounded above: For every $t \geq \max\{1, a\}$, we have $t^2 \geq a$, so such a t can no longer belong to A. By the completeness axiom, A therefore has a supremum; we set

$$x := \sup A.$$

We now show that $x^2 = a$. Suppose first that $x^2 < a$. Then also $(x + \delta)^2 < a$ for sufficiently small $\delta > 0$, so there would be an element of A that is larger than x. This contradicts the fact that x is the supremum of A.

Conversely, suppose that $x^2 > a$. Then one can show that there is a small $\delta > 0$ such that for all t with $x - \delta < t \leq x$, one already has $t^2 > a$. Then x would not be the least upper bound of A, since numbers just below x would already fail to be elements of A. This is again a contradiction.

Thus the only possibility is

$$x^2 = a.$$

Hence, for every $a > 0$, a square root exists in $\mathbb{R}$. $\square$

In particular, for $a = 2$ there also exists a real number x with $x^2 = 2$. Since $\sqrt{2}$ is not rational, it follows at the same time that the existence of square roots is, in general, a property of the real numbers, not of the rational numbers.

6.2 The Nested Interval Principle

The completeness axiom has several equivalent formulations. One particularly intuitive version is the so-called *nested interval principle.* Here one considers a sequence of nested closed intervals whose lengths become smaller and smaller. The completeness of the real numbers then guarantees that all these intervals have a common point.

Definition 156. *A sequence of closed intervals*

$$I_n = [a_n, b_n]$$

is called nested *if, for all* $n \in \mathbb{N}$,

$$I_{n+1} \subseteq I_n.$$

Intuitively, this means that each new interval lies entirely inside the preceding one. Thus the intervals are narrowed down further and further.

Theorem 157 (Nested interval principle). *Let*

$$I_n = [a_n, b_n]$$

be a sequence of nested closed intervals such that

$$\lim_{n\to\infty}(b_n - a_n) = 0.$$

Then there exists exactly one point

$$x \in \bigcap_{n=1}^{\infty} I_n.$$

Proof. Since the intervals are nested, we have

$$a_n \le a_{n+1} \le b_{n+1} \le b_n.$$

Thus the sequence (a_n) is monotonically increasing and bounded above, while (b_n) is monotonically decreasing and bounded below.

By the completeness of the real numbers, (a_n) has a limit

$$a := \lim_{n\to\infty} a_n,$$

and likewise (b_n) has a limit

$$b := \lim_{n\to\infty} b_n.$$

Since

$$0 \leq b_n - a_n \longrightarrow 0$$

it follows that

$$b - a = 0,$$

hence

$$a = b.$$

We denote this common value by x.

Since for every n we have

$$a_n \leq x \leq b_n,$$

x lies in every interval I_n. Therefore,

$$x \in \bigcap_{n=1}^{\infty} I_n.$$

Uniqueness follows immediately from the fact that the interval lengths converge to 0: two distinct points could not both lie in all sufficiently small intervals. □

The nested interval principle is equivalent to the completeness axiom of the real numbers. It provides a geometric perspective on completeness and will later be an important tool in the proof of the Bolzano–Weierstrass theorem.

6.2.1 Equivalence of the Completeness Axiom and the Nested Interval Principle

The completeness axiom and the nested interval principle are two different but equivalent ways of expressing the same fundamental idea. While the completeness axiom is formulated in terms of Cauchy sequences, the nested interval principle describes the same completeness property in geometric form. In what follows, we prove both directions of this equivalence.

Theorem 158. *The nested interval principle follows from the completeness axiom.*

Proof. Let

$$I_n = [a_n, b_n]$$

be a sequence of nested closed intervals such that

$$I_{n+1} \subseteq I_n \qquad \text{and} \qquad \lim_{n\to\infty}(b_n - a_n) = 0.$$

We consider the sequence of left endpoints (a_n).

Because the intervals are nested, for all $m \geq n$ we have

$$a_n \leq a_m \leq b_n.$$

In particular, it follows that

$$0 \leq a_m - a_n \leq b_n - a_n.$$

Since the interval lengths tend to 0, the sequence (a_n) is a Cauchy sequence: for every $\varepsilon > 0$, there exists an $N \in \mathbb{N}$ such that

$$b_N - a_N < \varepsilon.$$

Then, for all $m, n \geq N$, the nesting property implies

$$|a_m - a_n| \leq b_N - a_N < \varepsilon.$$

By the completeness axiom, every Cauchy sequence of real numbers converges. Hence there exists an $x \in \mathbb{R}$ such that

$$a_n \to x.$$

We now show that x lies in every interval I_n. To this end, fix an $n \in \mathbb{N}$. For all $m \geq n$, we have

$$a_n \leq a_m \leq b_n.$$

Letting $m \to \infty$, we obtain

$$a_n \leq x \leq b_n.$$

Thus $x \in I_n$.

Since this holds for every n, we obtain

$$x \in \bigcap_{n=1}^{\infty} I_n.$$

The uniqueness of the point of intersection follows from the fact that the interval lengths converge to 0: if another point y also belonged to all intervals, then for every n we would have

$$|x - y| \leq b_n - a_n.$$

Passing to the limit gives $|x - y| = 0$, and therefore $x = y$. □

Theorem 159. *The completeness axiom follows from the nested interval principle.*

Proof. Let (a_n) be a Cauchy sequence of real numbers. We want to show that (a_n) converges.

Since (a_n) is a Cauchy sequence, for every $k \in \mathbb{N}$ there exists an index n_k such that

$$|a_n - a_m| < 2^{-k} \qquad \text{for all } n, m \geq n_k.$$

We may choose these indices in such a way that

$$n_1 < n_2 < n_3 < \cdots$$

holds.

Now, for each $k \in \mathbb{N}$, we define the closed interval

$$I_k := \left[a_{n_k} - 2^{-k},\ a_{n_k} + 2^{-k}\right].$$

These intervals are nested. Indeed, let $x \in I_{k+1}$. Then

$$|x - a_{n_{k+1}}| \leq 2^{-(k+1)}.$$

Moreover, the Cauchy property gives

$$|a_{n_{k+1}} - a_{n_k}| < 2^{-(k+1)}.$$

Hence, by the triangle inequality,

$$|x - a_{n_k}| \leq |x - a_{n_{k+1}}| + |a_{n_{k+1}} - a_{n_k}| < 2^{-(k+1)} + 2^{-(k+1)} = 2^{-k}.$$

Thus $x \in I_k$, and therefore

$$I_{k+1} \subseteq I_k.$$

Furthermore, the lengths of the intervals tend to 0, since

$$\text{length}(I_k) = 2^{-k+1} \to 0.$$

By the nested interval principle, there therefore exists a point

$$x_0 \in \bigcap_{k=1}^{\infty} I_k.$$

We now show that $a_n \to x_0$. Let $\varepsilon > 0$ be given. Choose $k \in \mathbb{N}$ so large that

$$2^{-k+1} < \varepsilon.$$

For all $n \geq n_k$, the choice of n_k gives

$$|a_n - a_{n_k}| < 2^{-k}.$$

At the same time, since $x_0 \in I_k$, we have

$$|x_0 - a_{n_k}| \leq 2^{-k}.$$

Thus, by the triangle inequality,

$$|a_n - x_0| \leq |a_n - a_{n_k}| + |a_{n_k} - x_0| < 2^{-k} + 2^{-k} = 2^{-k+1} < \varepsilon.$$

This shows that

$$a_n \to x_0.$$

Hence every Cauchy sequence of real numbers converges.

□

Thus the completeness axiom and the nested interval principle are equivalent.

6.3 Subsequences

Not every sequence converges. Nevertheless, a sequence may still contain a convergent core. This is precisely the idea behind subsequences: instead of considering all terms of a sequence, one considers only a strictly increasing selection of indices.

Definition 160. *Let* $(a_n)_{n\in\mathbb{N}}$ *be a sequence of real numbers and let*

$$n_0 < n_1 < n_2 < \cdots$$

be a strictly increasing sequence of natural numbers. Then the sequence

$$(a_{n_k})_{k\in\mathbb{N}}$$

is called a subsequence *of* $(a_n)_{n\in\mathbb{N}}$.

Thus subsequences are obtained by selecting individual terms from a given sequence at increasing index positions. The original order is preserved, but some terms are omitted.

An important first connection is the following:

Theorem 161. *If a sequence* (a_n) *converges to a limit* a, *then every subsequence of* (a_n) *also converges to* a.

Proof. Let (a_{n_k}) be a subsequence of (a_n), and suppose that

$$\lim_{n\to\infty} a_n = a.$$

Then for every $\varepsilon > 0$ there exists an $N \in \mathbb{N}$ such that

$$|a_n - a| < \varepsilon \qquad \text{for all } n \geq N.$$

Since $n_k \to \infty$, there exists a $K \in \mathbb{N}$ such that $n_k \geq N$ for all $k \geq K$. Hence, for all $k \geq K$, we obtain

$$|a_{n_k} - a| < \varepsilon.$$

Thus the subsequence also converges to a. □

The interesting converse question is now the following: Can one still extract a convergent subsequence from a sequence that does not converge?

For bounded sequences of real numbers, the answer is surprisingly: yes.

Theorem 162 (Bolzano–Weierstrass). *Every bounded sequence of real numbers has a convergent subsequence.*

Proof. Let $(a_n)_{n\in\mathbb{N}}$ be a bounded sequence. Then there exist numbers $u, v \in \mathbb{R}$ such that

$$u \leq a_n \leq v \qquad \text{for all } n \in \mathbb{N}.$$

We set

$$I_0 := [u, v].$$

Thus this interval contains all terms of the sequence.

We now inductively construct a sequence of nested closed intervals

$$I_k = [u_k, v_k] \qquad (k \in \mathbb{N}),$$

such that each interval contains infinitely many terms of the sequence and the lengths of the intervals tend to 0.

Suppose that $I_k = [u_k, v_k]$ has already been constructed and contains infinitely many terms of the sequence. We divide I_k into two halves of equal length:

$$\left[u_k, \frac{u_k + v_k}{2}\right] \quad \text{and} \quad \left[\frac{u_k + v_k}{2}, v_k\right].$$

At least one of these two halves contains infinitely many terms of (a_n), since otherwise I_k could contain only finitely many terms. We call this half I_{k+1}. Then

$$I_{k+1} \subseteq I_k \qquad \text{and} \qquad \text{length}(I_{k+1}) = \frac{1}{2}\,\text{length}(I_k).$$

In particular,

$$\text{length}(I_k) = 2^{-k}\,\text{length}(I_0) \to 0.$$

We now choose a subsequence (a_{n_k}) such that

$$a_{n_k} \in I_k \qquad \text{for all } k \in \mathbb{N}.$$

This is possible because each I_k contains infinitely many terms of the sequence. Moreover, we can choose the indices to be strictly increasing, that is,

$$n_0 < n_1 < n_2 < \cdots.$$

By the nested interval principle, the sequence of nested intervals has a unique common point

$$x \in \bigcap_{k=0}^{\infty} I_k.$$

We now show that $a_{n_k} \to x$. Let $\varepsilon > 0$ be given. Choose $N \in \mathbb{N}$ so large that

$$\text{length}(I_N) < \varepsilon.$$

Then for all $k \geq N$ we have

$$a_{n_k} \in I_k \subseteq I_N \quad \text{and} \quad x \in I_N.$$

Hence both a_{n_k} and x lie in the same interval I_N, and therefore

$$|a_{n_k} - x| \leq \text{length}(I_N) < \varepsilon.$$

Thus we have shown that

$$a_{n_k} \to x.$$

Therefore every bounded sequence of real numbers has a convergent subsequence. □

Definition 163. *A number $x \in \mathbb{R}$ is called an* accumulation point *or* limit of a subsequence *of the sequence (a_n) if there exists a subsequence (a_{n_k}) such that*

$$a_{n_k} \to x.$$

The Bolzano–Weierstrass theorem can therefore also be formulated as follows:

Theorem 164. *Every bounded sequence of real numbers has at least one accumulation point.*

This point of view is particularly important for further analysis, because it makes it possible to “extract” limit values from a single bounded sequence, even if the sequence as a whole does not converge.

6.4 Exercises and solutions

Exercise 165. *Determine whether the following sequences converge. Where applicable, find all accumulation points and construct convergent subsequences.*

1. $a_n = (-1)^n$
2. $a_n = \sin(n\pi)$
3. $a_n = (-1)^n + \frac{1}{n}$
4. $a_n = \cos\left(\frac{n\pi}{2}\right)$

Solution 166. 1. *For* $a_n = (-1)^n$, *we have:*

$$a_{2k} = 1, \qquad a_{2k+1} = -1.$$

Thus the sequence itself does not converge, since it alternates between two values. However, there are convergent subsequences:

$$\begin{aligned}(a_{2k}) &= (1, 1, 1, \dots) \ \to 1, \\ (a_{2k+1}) &= (-1, -1, -1, \dots) \to -1.\end{aligned}$$

Therefore the only accumulation points are 1 *and* -1.

2. *Since* $n\pi$ *is an integer multiple of* π *for every* $n \in \mathbb{N}$, *we have*

$$\sin(n\pi) = 0 \qquad \textit{for all } n.$$

The sequence is therefore constant and hence convergent:

$$a_n \to 0.$$

The only accumulation point is 0.

3. *For*

$$a_n = (-1)^n + \frac{1}{n}$$

we again consider even and odd indices:

$$a_{2k} = 1+\frac{1}{2k} \to 1, \qquad a_{2k+1} = -1+\frac{1}{2k+1} \to -1.$$

The sequence itself does not converge, but it has two convergent subsequences. Its accumulation points are precisely 1 *and* −1.

4. *The sequence*

$$a_n = \cos\left(\frac{n\pi}{2}\right)$$

is periodic with values

$$1,\ 0,\ -1,\ 0,\ 1,\ 0,\ -1,\ 0, \ldots$$

Therefore it does not converge. However, there are several convergent subsequences, for example

$$a_{4k} = 1, \qquad a_{4k+1} = 0,$$
$$a_{4k+2} = -1, \qquad a_{4k+3} = 0.$$

Thus the accumulation points are

$$-1,\ 0,\ 1.$$

Exercise 167. *Show that if a sequence* (a_n) *is convergent, then it has exactly one accumulation point.*

Solution 168. *Let*

$$a_n \to a.$$

Then a *is an accumulation point, since the sequence itself is a subsequence of itself.*

Now let b *be any accumulation point of the sequence. Then there exists a subsequence* (a_{n_k}) *such that*

$$a_{n_k} \to b.$$

Since every subsequence of a convergent sequence converges to the same limit, we also have

$$a_{n_k} \to a.$$

Limits are unique; hence

$$b = a$$

must hold.

Thus the sequence has exactly one accumulation point, namely its limit.

Exercise 169. *Give an example of*

1. *an unbounded sequence without a convergent subsequence,*

2. *a bounded sequence with exactly two accumulation points,*

3. *a bounded sequence with infinitely many accumulation points.*

Solution 170. 1. *A simple example is*

$$a_n = n.$$

This sequence is unbounded. Every subsequence is also unbounded and therefore cannot converge.

2. *An example is*

$$a_n = (-1)^n.$$

This sequence is bounded and has exactly the two accumulation points 1 *and* −1.

3. *A possible example is a sequence that lists the numbers*

$$1, \frac{1}{2}, \frac{1}{3}, \frac{1}{4}, \dots$$

with repetitions in such a way that each number $\frac{1}{m}$ *occurs infinitely often. For instance, one may take the sequence*

$$\begin{aligned} a_1 &= 1, & a_2 &= \frac{1}{2}, & a_3 &= 1, & a_4 &= \frac{1}{3}, \\ a_5 &= 1, & a_6 &= \frac{1}{2}, & a_7 &= 1, & a_8 &= \frac{1}{4}, \\ a_9 &= 1, & a_{10} &= \frac{1}{2}, & a_{11} &= 1, & a_{12} &= \frac{1}{3}, \dots \end{aligned}$$

More explicitly, one can arrange the terms in blocks:

$$1; \quad 1, \frac{1}{2}; \quad 1, \frac{1}{2}, \frac{1}{3}; \quad 1, \frac{1}{2}, \frac{1}{3}, \frac{1}{4}; \quad \dots$$

Then the sequence is bounded, since all its terms lie in $[0, 1]$*. Each number* $\frac{1}{m}$ *occurs infinitely often and is therefore an accumulation point. In addition,* 0 *is an accumulation point, because* $\frac{1}{m} \to 0$ *as* $m \to \infty$*. Thus the sequence has infinitely many accumulation points.*

Exercise 171. *Show that if a sequence is bounded and has exactly one accumulation point, then the sequence converges to this accumulation point.*

Solution 172. *Let* a *be the unique accumulation point of the bounded sequence* (a_n)*.*

We show that $a_n \to a$*. Suppose, on the contrary, that this is not the case. Then there would exist an* $\varepsilon > 0$ *and infinitely many indices* n *such that*

$$|a_n - a| \geq \varepsilon.$$

From these terms one can choose a subsequence (a_{n_k}) *such that, for all* k*,*

$$|a_{n_k} - a| \geq \varepsilon.$$

This subsequence is again bounded, and therefore, by the Bolzano–Weierstrass theorem, it has a convergent subsequence. The limit of that subsequence would then be an accumulation point of the original sequence. Because of the estimate

$$|a_{n_k} - a| \geq \varepsilon$$

however, this limit cannot be a*. This contradicts the assumption that* a *is the unique accumulation point.*

Hence

$$a_n \to a$$

must hold.

Exercise 173. *Let* (a_n) *be a bounded sequence, and let* (a_{n_k}) *be a convergent subsequence with limit* a. *Show that every subsequence of* (a_{n_k}) *also converges to* a.

Solution 174. *The subsequence* (a_{n_k}) *converges to* a. *Every subsequence of a convergent sequence again converges to the same limit.*

Therefore, for every subsequence $(a_{n_{k_j}})$, *we have*

$$a_{n_{k_j}} \to a.$$

Intuitively, this can also be understood as follows: if a sequence is already sufficiently close to a limit, then any further selected subsequence remains close to this limit as well.

Exercise 175. *Consider the sequence*

$$a_n = \frac{(-1)^n n}{n+1}.$$

1. *Show that* (a_n) *is bounded.*

2. *Determine two different convergent subsequences.*

3. *Determine all accumulation points of the sequence.*

Solution 176. *1. We have*

$$|a_n| = \frac{n}{n+1} < 1$$

for all $n \in \mathbb{N}$. Hence the sequence is bounded.

2. For the even indices, we obtain

$$a_{2k} = \frac{2k}{2k+1} \to 1.$$

For the odd indices, we obtain

$$a_{2k+1} = -\frac{2k+1}{2k+2} \to -1.$$

Thus we have found two different convergent subsequences.

3. From the two subsequences found above, we immediately see that 1 and -1 are accumulation points. There are no further accumulation points, since every subsequence of (a_n) contains either infinitely many even indices, infinitely many odd indices, or a mixture of both. However, any convergent subsequence must eventually reduce to one of these two cases, and then its limit is either 1 or -1.

Thus the only accumulation points are

$$-1 \quad \textit{and} \quad 1.$$

6.5 Monotonicity of Sequences

Besides boundedness, the monotonicity of sequences also plays a central role in analysis. Monotone sequences change in only one direction: they either keep increasing or keep decreasing. Together with boundedness, this leads to a fundamental convergence criterion.

Definition 177. *A sequence $(a_n)_{n\in\mathbb{N}}$ is called*

1. monotonically increasing, *if*
$$a_n \le a_{n+1} \qquad \textit{for all } n \in \mathbb{N},$$
2. strictly monotonically increasing, *if*
$$a_n < a_{n+1} \qquad \textit{for all } n \in \mathbb{N},$$
3. monotonically decreasing, *if*
$$a_n \ge a_{n+1} \qquad \textit{for all } n \in \mathbb{N},$$
4. strictly monotonically decreasing, *if*
$$a_n > a_{n+1} \qquad \textit{for all } n \in \mathbb{N}.$$

Monotonicity alone is not yet sufficient for convergence. The sequence

$$a_n = n$$

is, for example, strictly monotonically increasing, but unbounded and therefore not convergent. Only the interplay between monotonicity and boundedness yields the following important result.

Theorem 178 (Monotone convergence criterion). *Every bounded monotone sequence of real numbers converges.*

Proof. Let (a_n) be a bounded monotone sequence. By the Bolzano–Weierstrass theorem, (a_n) has a convergent subsequence

$$(a_{n_k}).$$

Let

$$a_{n_k} \to a.$$

We now show that the entire sequence already converges to a.

We first consider the case where (a_n) is monotonically increasing. The case of monotonically decreasing sequences is treated analogously.

Let $\varepsilon > 0$ be given. Since

$$a_{n_k} \to a,$$

there exists a $k_0 \in \mathbb{N}$ such that

$$|a_{n_k} - a| < \varepsilon \qquad \text{for all } k \geq k_0.$$

Set

$$N := n_{k_0}.$$

For every $n \geq N$, there is a $k \geq k_0$ such that

$$n_k \leq n < n_{k+1}.$$

Since the sequence is monotonically increasing, it follows that

$$a_{n_k} \leq a_n \leq a_{n_{k+1}}.$$

Moreover, both a_{n_k} and $a_{n_{k+1}}$ converge to a. Thus, for sufficiently large k, we have

$$a - \varepsilon < a_{n_k} \leq a_n \leq a_{n_{k+1}} < a + \varepsilon.$$

Hence

$$|a_n - a| < \varepsilon.$$

Since this holds for all $n \geq N$, we obtain

$$a_n \to a.$$

Therefore every bounded monotonically increasing sequence converges. The monotonically decreasing case is analogous. □

The monotone convergence criterion is one of the most important convergence criteria in analysis. It shows that a sequence already converges if it moves only in one direction and remains bounded.

6.6 Exercises and Solutions

Exercise 179. *Examine the sequence*

$$a_n = \frac{n}{n+1}.$$

1. *Is the sequence monotone?*

2. *Is it bounded?*

3. *Deduce its convergence and determine its limit.*

Solution 180. 1. *We consider the difference of consecutive terms:*

$$a_{n+1} - a_n = \frac{n+1}{n+2} - \frac{n}{n+1} = \frac{(n+1)^2 - n(n+2)}{(n+1)(n+2)} = \frac{1}{(n+1)(n+2)} > 0.$$

Thus (a_n) is strictly monotonically increasing.

2. *Moreover, for all n we have:*

$$0 < a_n = \frac{n}{n+1} < 1.$$

Hence the sequence is bounded.

3. *By the monotone convergence criterion, every bounded monotone sequence converges. Therefore (a_n) has a limit. Since*

$$a_n = \frac{n}{n+1} = 1 - \frac{1}{n+1},$$

it follows immediately that

$$a_n \to 1.$$

Exercise 181. *Examine the sequence*

$$b_n = \sqrt{n+1} - \sqrt{n}.$$

1. *Is the sequence monotone?*

2. *Is it bounded?*

3. *Deduce its convergence and determine its limit.*

Solution 182. 1. *First, we write the sequence in rationalized form:*

$$b_n = \sqrt{n+1} - \sqrt{n} = \frac{(n+1)-n}{\sqrt{n+1}+\sqrt{n}} = \frac{1}{\sqrt{n+1}+\sqrt{n}}.$$

Since the denominator increases as n increases, the fraction becomes smaller. Thus (b_n) is monotonically decreasing.

2. *Moreover, for all n we have:*

$$b_n > 0.$$

Hence the sequence is bounded below by 0.

3. *By the monotone convergence criterion, (b_n) converges. To determine the limit, we again use the representation*

$$b_n = \frac{1}{\sqrt{n+1}+\sqrt{n}}.$$

The denominator tends to ∞, and therefore

$$b_n \to 0.$$

6.7 The Axioms of the Real Numbers

The real numbers are distinguished by the fact that they possess several fundamental structural properties at the same time. These properties can be summarized in four axioms:

- *Field axioms:* The usual rules of arithmetic for addition and multiplication hold, in particular associativity, commutativity, distributivity, and the existence of neutral and inverse elements.

- *Order axioms:* The real numbers are ordered, and this order is compatible with addition and multiplication.

- *Archimedean axiom:* For every positive real number x, there exists a natural number n such that

$$n > x.$$

  Intuitively, this means that there are no infinitely large or infinitely small real numbers.

- *Completeness axiom:* Every Cauchy sequence of real numbers converges in $\mathbb{R}$.

A field that satisfies these four properties is called a *complete Archimedean ordered field.* This terminology captures exactly the structure needed in analysis.

Theorem 183. *Every complete Archimedean ordered field is, up to a unique isomorphism, the field of real numbers.*

Proof sketch. The Archimedean axiom ensures that the field can be compared with the rational numbers: the natural numbers, and hence also the rational numbers, can be embedded into it in a unique way. Moreover, every such field can be densely approximated by rational numbers.

Completeness then shows that every Cauchy sequence of rational approximations has a limit and that this limit is unique. Thus the field is "without gaps" in exactly the same sense as $\mathbb{R}$.

Since the real numbers themselves form a complete Archimedean ordered field, and since such a field is uniquely determined by its rational approximations, it follows that all complete Archimedean ordered fields are isomorphic to one another. In particular, they are isomorphic to $\mathbb{R}$. □

This characterizes the structure of the real numbers as the number system of analysis. In the next step, we can use these properties to prove the existence of square roots of positive real numbers.

6.7.1 Outlook

The results obtained so far show that the real numbers are not only an ordered field, but are also characterized by a principle of completeness. It is precisely this completeness that makes it possible to answer many existence questions positively which remain unresolved within the rational numbers.

A first central example is the question of square roots. In $\mathbb{Q}$, not every positive number is the square of a rational number; in particular, $\sqrt{2}$ is not rational. In $\mathbb{R}$, by contrast, completeness is sufficient to show that every positive number has a square root.

Thus we obtain an important first existence theorem of analysis: The real numbers are exactly large enough to guarantee solutions of such equations within the number system itself. In the next section, we therefore show that for every number $a > 0$ there exists a real number x such that

$$x^2 = a$$

holds.

Chapter 7

Existence of Square Roots

In the previous chapter, we introduced the completeness axiom. It states that every nonempty subset of $\mathbb{R}$ that is bounded above has a least upper bound. This axiom reveals its strength precisely in situations where numbers can no longer be described directly, but only as limits or bounds. A first fundamental example is the existence of square roots.

In what follows, let $a \geq 0$ be fixed. For $a = 0$, the situation is clear, since

$$0^2 = 0.$$

Thus the interesting case is $a > 0$.

7.1 Supremum and Infimum

Before constructing the square root, we introduce two fundamental notions.

Theorem 184 (Supremum and Infimum). *Let $M \subseteq \mathbb{R}$ be a nonempty set.*

- *A number $u \in \mathbb{R}$ is called an* upper bound *of M if, for all $x \in M$,*
$$x \leq u.$$
- *If M is bounded above, then the least upper bound of M is called the* supremum *of M, and one writes*
$$\sup M.$$
- *A number $l \in \mathbb{R}$ is called a* lower bound *of M if, for all $x \in M$,*
$$l \leq x.$$
- *If M is bounded below, then the greatest lower bound of M is called the* infimum *of M, and one writes*
$$\inf M.$$

Thus the completeness axiom says: Every nonempty subset of $\mathbb{R}$ that is bounded above has a supremum. Dually, every nonempty subset of $\mathbb{R}$ that is bounded below has an infimum.

7.2 The Existence of a Positive Square Root

Theorem 185. *For every number $a > 0$, there exists exactly one positive real number s such that*

$$s^2 = a.$$

Proof. We consider the set

$$S_a := \{x \in \mathbb{R}_{\geq 0} \mid x^2 < a\}.$$

The set S_a is nonempty, since $0 \in S_a$. Moreover, it is bounded above: indeed, if $x > a + 1$, then

$$x^2 > (a+1)^2 > a.$$

Hence, by the completeness axiom, S_a has a supremum. We set

$$s := \sup S_a.$$

We now show that

$$s^2 = a$$

must hold.

First, s^2 cannot be less than a. Suppose that

$$s^2 < a.$$

Then there exists a sufficiently small $\varepsilon > 0$ such that

$$(s + \varepsilon)^2 < a.$$

But then we would also have

$$s + \varepsilon \in S_a,$$

which contradicts the fact that s is an upper bound of S_a.

Thus
$$s^2 \not< a.$$

Similarly, s^2 cannot be greater than a. Suppose that

$$s^2 > a.$$

Then we can choose $\varepsilon > 0$ so small that

$$s - \varepsilon > 0 \qquad \text{and} \qquad (s - \varepsilon)^2 > a.$$

Then, for every $x \in S_a$, the following holds: from

$$x \geq s - \varepsilon$$

it would follow, by the monotonicity of the square function on $\mathbb{R}_{\geq 0}$, that
$$x^2 \geq (s - \varepsilon)^2 > a,$$

and hence
$$x \notin S_a.$$

Therefore, for all $x \in S_a$, we have

$$x < s - \varepsilon.$$

Thus $s - \varepsilon$ is an upper bound of S_a. This contradicts the fact that s is the least upper bound of S_a.

Thus also
$$s^2 \not> a.$$

Since s^2 can be neither less than nor greater than a, the only possibility is
$$s^2 = a.$$

It remains to prove uniqueness. Let $u > 0$ and $v > 0$ satisfy
$$u^2 = a = v^2.$$
Then it follows that
$$u^2 - v^2 = 0.$$
Hence
$$(u - v)(u + v) = 0.$$
Since
$$u + v > 0,$$
we must have
$$u - v = 0$$
and therefore
$$u = v.$$

Thus, for every number $a > 0$, there exists exactly one positive real number whose square is a. □

Definition 186. *The uniquely determined positive number s satisfying*
$$s^2 = a$$

is called the square root *of a and is denoted by*

$$\sqrt{a}$$

.

7.3 Heron's Method

From the existence of the square root, we now obtain an effective approximation method.

The basic idea is simple: If x is already an approximation of $\sqrt{a}$, then

$$\frac{a}{x}$$

is also an approximation. The arithmetic mean of these two numbers is usually closer to the desired root.

This leads to the recursion

$$x_{n+1} := \frac{1}{2}\left(x_n + \frac{a}{x_n}\right) \qquad (n \geq 0).$$

As an initial value, we may choose, for example,

$$x_0 := \max\{1, a\}.$$

Theorem 187. *The sequence (x_n) has the properties*

$$x_n > 0, \qquad x_n^2 \geq a, \qquad x_{n+1} \leq x_n$$

for all $n \geq 0$.

Proof. The initial value satisfies the assertion.

Assume that the statements hold for some n. Then

$$x_n > 0.$$

Consequently, we also have

$$\frac{a}{x_n} > 0,$$

and hence

$$x_{n+1} > 0.$$

Moreover,

$$x_{n+1}^2 - a = \left(\frac{1}{2}\left(x_n + \frac{a}{x_n}\right)\right)^2 - a.$$

Expanding gives

$$x_{n+1}^2 - a = \frac{(x_n^2 - a)^2}{4x_n^2}.$$

Since squares are nonnegative, it follows that

$$x_{n+1}^2 - a \geq 0.$$

Thus

$$x_{n+1}^2 \geq a.$$

From

$$x_n^2 \geq a$$

and $x_n > 0$, we obtain

$$\frac{a}{x_n} \leq x_n.$$

Therefore,

$$x_{n+1} = \frac{1}{2}\left(x_n + \frac{a}{x_n}\right) \leq \frac{1}{2}(x_n + x_n) = x_n.$$

This proves all assertions. □

Theorem 188. *The sequence (x_n) converges to $\sqrt{a}$.*

Proof. By the preceding theorem, the sequence is monotonically decreasing and bounded below by 0. Hence, by the completeness axiom, it has a limit:

$$x_n \to x \qquad (n \to \infty)$$

for some $x \geq 0$.

Since the preceding theorem also gives $x_n^2 \geq a$ for all n, we have $x \geq \sqrt{a} > 0$. Therefore,

$$\frac{a}{x_n} \to \frac{a}{x}.$$

We now pass to the limit in the recursion formula

$$x_{n+1} = \frac{1}{2}\left(x_n + \frac{a}{x_n}\right)$$

and obtain

$$x = \frac{1}{2}\left(x + \frac{a}{x}\right).$$

Multiplying by $2x$ gives

$$2x^2 = x^2 + a.$$

Thus,

$$x^2 = a.$$

Since $\sqrt{a}$ is the unique positive solution of this equation, it follows that

$$x = \sqrt{a}.$$

Consequently,

$$\lim_{n\to\infty} x_n = \sqrt{a}.$$

□

7.4 Quadratic Convergence

Theorem 189. *Let*

$$r := \sqrt{a}.$$

Then

$$x_{n+1} - r = \frac{(x_n - r)^2}{2x_n}.$$

Proof. Since

$$r^2 = a,$$

we have

$$x_{n+1} - r = \frac{1}{2}\left(x_n + \frac{a}{x_n}\right) - r = \frac{x_n^2 + a - 2rx_n}{2x_n}.$$

Using

$$a = r^2$$

it follows that

$$x_{n+1} - r = \frac{x_n^2 - 2rx_n + r^2}{2x_n} = \frac{(x_n - r)^2}{2x_n}.$$

□

Thus, the new error is essentially the square of the previous error. For this reason, one speaks of *quadratic convergence.*

If x_n is already close to $\sqrt{a}$, then the next step improves the approximation exceptionally strongly. For this reason, Heron's method is one of the classical and practically very efficient methods for computing square roots.

Summary 190. *The completeness axiom implies the existence of suprema and infima. This makes it possible to show that for every number* $a \geq 0$ *there exists exactly one positive number* $\sqrt{a}$ *such that*

$$(\sqrt{a})^2 = a$$

holds.

Moreover, the recursion

$$x_{n+1} = \frac{1}{2}\left(x_n + \frac{a}{x_n}\right)$$

provides a simple and very rapidly converging method for computing square roots.

Chapter 8

Convergence Criteria for Series

In the preceding chapters, we used the completeness of the real numbers primarily for sequences. There we saw that a sequence converges if and only if, from some index onward, its terms become arbitrarily close to one another; in other words, if and only if it is a Cauchy sequence.

This idea can be transferred directly to infinite series. Indeed, a series

$$\sum_{n=0}^{\infty} a_n$$

is, at its core, nothing other than a sequence of partial sums

$$S_n = \sum_{k=0}^{n} a_k.$$

Thus the question whether a series converges is equivalent to the question whether the sequence of its partial sums converges. This is precisely where the completeness axiom comes into play again: in the real numbers, exactly the Cauchy sequences converge.

For series, this gives us a first fundamental convergence criterion. In practice, it is often especially useful because it does not require knowing the limits of the partial sums, but only requires control over the "tails" of the series.

8.1 Cauchy Convergence Criterion for Series

Theorem 191 (Cauchy convergence criterion for series). *Let (a_n) be a sequence of real numbers. Then the series*

$$\sum_{n=0}^{\infty} a_n$$

converges if and only if for every $\epsilon > 0$ there exists an $N \in \mathbb{N}$ such that

$$\left| \sum_{k=m}^{n} a_k \right| < \epsilon \qquad \textit{for all } n \geq m \geq N.$$

Proof. We denote by

$$S_n := \sum_{k=0}^{n} a_k$$

the n-th partial sum of the series. Then the series $\sum_{n=0}^{\infty} a_n$ is convergent if and only if the sequence (S_n) converges.

By the Cauchy criterion for sequences, it is therefore enough to show that (S_n) is a Cauchy sequence if and only if the estimate above holds.

First suppose that (S_n), and hence the series, is convergent. Then (S_n) is in particular a Cauchy sequence. Thus, for every $\epsilon > 0$, there exists an $N \in \mathbb{N}$ such that

$$|S_n - S_{m-1}| < \epsilon \qquad \text{for all } n \geq m \geq N.$$

But since

$$S_n - S_{m-1} = \sum_{k=0}^{n} a_k - \sum_{k=0}^{m-1} a_k = \sum_{k=m}^{n} a_k,$$

it follows immediately that

$$\left| \sum_{k=m}^{n} a_k \right| < \epsilon \qquad \text{for all } n \geq m \geq N.$$

Conversely, suppose that the stated condition holds. Then for $n \geq m \geq N$ we may write

$$S_n - S_{m-1} = \sum_{k=m}^{n} a_k,$$

and hence

$$|S_n - S_{m-1}| < \epsilon.$$

Thus (S_n) is a Cauchy sequence. By the completeness of the real numbers, (S_n) converges, and therefore the series $\sum_{n=0}^{\infty} a_n$ converges as well.

This proves the equivalence. □

Theorem 192 (Necessary condition for the convergence of series). *A necessary, but in general not sufficient, condition for the convergence of the series*

$$\sum_{n=0}^{\infty} a_n$$

is

$$\lim_{n\to\infty} a_n = 0.$$

Proof. Assume that the series

$$\sum_{n=0}^{\infty} a_n$$

converges. Then, by the Cauchy convergence criterion for series, the following holds:

For every $\epsilon > 0$ there exists an $N \in \mathbb{N}$ such that

$$\left|\sum_{k=m}^{n} a_k\right| < \epsilon \qquad \text{for all } n \geq m \geq N.$$

Now choose, in this estimate, the special case $m = n$. Then, for all $n \geq N$, it follows that

$$\left|\sum_{k=n}^{n} a_k\right| < \epsilon.$$

But since

$$\sum_{k=n}^{n} a_k = a_n,$$

we obtain

$$|a_n| < \epsilon \qquad \text{for all } n \geq N.$$

This is precisely the definition of

$$\lim_{n\to\infty} a_n = 0.$$

Thus $a_n \to 0$ is a necessary condition for the convergence of the series.

However, this condition is in general not sufficient. There are series with $a_n \to 0$ that nevertheless diverge, for example the harmonic series

$$\sum_{n=1}^{\infty} \frac{1}{n}.$$

□

Remark 193. *The statement $a_n \to 0$ is therefore only a first necessary test: if it fails, the series cannot converge. If it holds, however, nothing has yet been decided about convergence.*

Theorem 194 (Convergence of monotonically increasing series)**.** *Let (a_n) be a sequence of real numbers with $a_n \geq 0$ for all $n \in \mathbb{N}$. Then the series*

$$\sum_{n=0}^{\infty} a_n$$

converges if and only if the sequence of its partial sums

$$s_n := \sum_{k=0}^{n} a_k$$

is bounded.

Proof. We consider the sequence of partial sums

$$s_n = \sum_{k=0}^{n} a_k.$$

Since all terms a_n are nonnegative, we have

$$s_{n+1} = s_n + a_{n+1} \geq s_n$$

for all $n \in \mathbb{N}$. Hence the sequence (s_n) is monotonically increasing.

$\Rightarrow$ If the series $\sum_{n=0}^{\infty} a_n$ converges, then the sequence (s_n) of partial sums converges. Every convergent sequence is bounded, so (s_n) is bounded.

$\Leftarrow$ If (s_n) is bounded, then (s_n) is a monotonically increasing and bounded sequence. By the theorem on the convergence of bounded monotone sequences, (s_n) converges. Therefore the series $\sum_{n=0}^{\infty} a_n$ also converges.

Thus the series is convergent if and only if its sequence of partial sums is bounded. □

Remark 195. *This criterion is particularly important because, for series with nonnegative terms, it reduces the question of convergence to a simple estimate of the partial sums.*

8.2 The Geometric Series

The geometric series is one of the most important examples in analysis. Many later developments build on it, such as power series, Taylor series, and Fourier expansions.

A geometric series arises when each term is obtained from the preceding one by multiplication with a fixed number q:

$$1,\ q,\ q^2,\ q^3,\ \ldots$$

The corresponding series is

$$\sum_{n=0}^{\infty} q^n.$$

The decisive question is now: For which values of q does this series converge?

We first compute the partial sums.

Lemma 196. *For $q \neq 1$, one has*

$$\sum_{k=0}^{n} q^k = \frac{1-q^{n+1}}{1-q}.$$

Proof. We set

$$s_n := \sum_{k=0}^{n} q^k = 1 + q + q^2 + \cdots + q^n.$$

Multiplication by q gives

$$qs_n = q + q^2 + \cdots + q^n + q^{n+1}.$$

Subtracting the two equations, almost all terms cancel:

$$s_n - qs_n = 1 - q^{n+1}.$$

Thus

$$(1-q)s_n = 1 - q^{n+1}.$$

Since $q \neq 1$, we may divide by $1-q$ and obtain

$$s_n = \frac{1-q^{n+1}}{1-q}.$$

□

With this formula, we can now describe the convergence behavior completely.

Theorem 197 (Convergence of the geometric series). *The geometric series*

$$\sum_{n=0}^{\infty} q^n$$

converges if and only if

$$|q| < 1.$$

In this case,

$$\sum_{n=0}^{\infty} q^n = \frac{1}{1-q}.$$

Proof. For $q = 1$, one obtains

$$1 + 1 + 1 + \ldots,$$

which is a divergent series.

Now let $q \neq 1$. By the preceding lemma, the n-th partial sum satisfies

$$s_n = \frac{1 - q^{n+1}}{1 - q}.$$

$\Rightarrow$ First suppose that $|q| < 1$. Then

$$q^{n+1} \to 0 \qquad (n \to \infty),$$

and hence

$$s_n = \frac{1 - q^{n+1}}{1 - q} \to \frac{1}{1 - q}.$$

Thus the geometric series converges, and its sum is

$$\frac{1}{1 - q}.$$

$\Leftarrow$ Conversely, if $|q| > 1$, then $|q|^{n+1}$ grows without bound. Therefore the sequence

$$\frac{1 - q^{n+1}}{1 - q}$$

has no finite limit. Hence the series diverges.

It remains to consider the case $q = -1$. Then one obtains

$$1 - 1 + 1 - 1 + \ldots.$$

The partial sums are

$$1, 0, 1, 0, \dots$$

and have no limit. Thus the series also diverges in this case.

Consequently, the geometric series converges exactly for $|q| < 1$. □

Example 198. *For $q = \frac{1}{2}$, one obtains*

$$1 + \frac{1}{2} + \frac{1}{4} + \frac{1}{8} + \cdots = \frac{1}{1 - \frac{1}{2}} = 2.$$

For $q = \frac{1}{3}$, correspondingly,

$$1 + \frac{1}{3} + \frac{1}{9} + \frac{1}{27} + \cdots = \frac{1}{1 - \frac{1}{3}} = \frac{3}{2}.$$

Remark 199. *The geometric series illustrates impressively that infinitely many positive summands can nevertheless have a finite sum.*

8.3 Convergence of Alternating Series

In addition to series with only nonnegative terms, series with alternating signs also play a central role. Such series are called *alternating series.* A typical example is the series

$$1 - \frac{1}{2} + \frac{1}{3} - \frac{1}{4} + \cdots.$$

In alternating series, positive and negative contributions often partially cancel each other out. As a result, convergence can occur even when the absolute values of the terms tend to 0 only slowly.

An important and particularly useful criterion for such series is the Leibniz convergence criterion.

Theorem 200 (Leibniz convergence criterion). *Let (a_n) be a monotonically decreasing sequence of nonnegative numbers such that*

$$\lim_{n\to\infty} a_n = 0,$$

then the alternating series

$$\sum_{n=0}^{\infty}(-1)^n a_n$$

converges.

Proof. We consider the sequence of partial sums

$$s_n := \sum_{k=0}^{n}(-1)^k a_k.$$

To study the convergence of (s_n), we separate the even and odd partial sums.

We have

$$s_{2n} = (a_0 - a_1) + (a_2 - a_3) + \cdots + (a_{2n-2} - a_{2n-1}) + a_{2n},$$

and

$$s_{2n+1} = (a_0 - a_1) + (a_2 - a_3) + \cdots + (a_{2n} - a_{2n+1}).$$

Since (a_n) is monotonically decreasing, for all n we have

$$a_{2n} - a_{2n+1} \geq 0.$$

It follows that

$$s_{2n+2} - s_{2n} = a_{2n+2} - a_{2n+1} \leq 0,$$

so the sequence (s_{2n}) is monotonically decreasing.

Moreover,

$$s_{2n+1} - s_{2n} = -a_{2n+1} \leq 0,$$

so s_{2n+1} always lies below s_{2n}. On the other hand, we obtain

$$s_{2n+1} = s_{2n} - a_{2n+1},$$

and since $a_{2n+1} \geq 0$, the odd partial sums do not exceed the even partial sums.

Even more important is the relation between consecutive even and odd partial sums:

$$s_{2n} - s_{2n+1} = a_{2n+1}.$$

Since $a_{2n+1} \to 0$, the distances between the even and odd partial sums become arbitrarily small.

Now consider the two sequences (s_{2n}) and (s_{2n+1}). The sequence (s_{2n}) is monotonically decreasing and bounded below by the sequence (s_{2n+1}), hence it converges. The sequence (s_{2n+1}) is monotonically increasing and bounded above by (s_{2n}), hence it also converges.

Let

$$\lim_{n\to\infty} s_{2n} = s \qquad \text{and} \qquad \lim_{n\to\infty} s_{2n+1} = t.$$

From

$$s_{2n} - s_{2n+1} = a_{2n+1}$$

and $a_{2n+1} \to 0$, taking limits gives

$$s - t = 0,$$

and therefore $s = t$.

Thus the even and odd partial sums have the same limit. Consequently, the entire sequence (s_n) converges, and hence the series

$$\sum_{n=0}^{\infty} (-1)^n a_n$$

converges. □

Remark 201. *The Leibniz criterion shows that, for alternating series, it is not only the size of the terms that matters, but also the change of sign. Even if the absolute values a_n tend to 0 only slowly, the series may still converge.*

Theorem 202 (Estimate of the remainder for alternating series)**.** *Let (a_n) be a monotonically decreasing sequence of nonnegative numbers with*

$$\lim_{n\to\infty} a_n = 0.$$

Then the alternating series

$$\sum_{n=0}^{\infty}(-1)^n a_n$$

converges. If

$$s := \sum_{n=0}^{\infty}(-1)^n a_n$$

denotes its sum and

$$s_n := \sum_{k=0}^{n}(-1)^k a_k$$

denotes the n-th partial sum, then the remainder

$$R_n := s - s_n$$

satisfies the estimate

$$|R_n| \leq a_{n+1}.$$

Proof. We use the fact, already observed in the proof of the Leibniz convergence criterion, that the even and odd partial sums form two monotone sequences converging to the same limit s.

For the even and odd partial sums, we have

$$s_{2n} \geq s \geq s_{2n+1}.$$

This is because the even partial sums lie above the limit, while the odd partial sums lie below it.

We now consider two cases.

If n is even, say $n = 2m$, then

$$s_{2m+1} \leq s \leq s_{2m}.$$

Thus s is enclosed between two consecutive partial sums. Hence

$$|s - s_{2m}| \leq s_{2m} - s_{2m+1} = a_{2m+1}.$$

If n is odd, say $n = 2m+1$, then correspondingly

$$s_{2m+1} \leq s \leq s_{2m+2},$$

and therefore

$$|s - s_{2m+1}| \leq s_{2m+2} - s_{2m+1} = a_{2m+2}.$$

In both cases, the remainder is therefore bounded by the absolute value of the first term not yet included:

$$|R_n| = |s - s_n| \leq a_{n+1}.$$

□

Remark 203. *This estimate is particularly useful because it provides a very simple error control: if one wants to determine the sum of an alternating series up to an error smaller than ϵ, it is enough to add terms until*

$$a_{n+1} < \epsilon.$$

Example 204 (A simple example). *Consider the series*

$$\sum_{n=0}^{\infty} \frac{(-1)^n}{n+1} = 1 - \frac{1}{2} + \frac{1}{3} - \frac{1}{4} + \cdots.$$

Here

$$a_n = \frac{1}{n+1}.$$

The sequence (a_n) is monotonically decreasing, nonnegative, and satisfies

$$\lim_{n\to\infty} a_n = 0.$$

Therefore, the series converges by the Leibniz convergence criterion.

Let us compute a few partial sums:

$$s_0 = 1, \qquad s_1 = \frac{1}{2}, \qquad s_2 = \frac{5}{6}, \qquad s_3 = \frac{7}{12}, \qquad s_4 = \frac{47}{60}.$$

One can see that the partial sums approach a limit alternately from above and below. The series converges, even though its terms tend to 0 only slowly.

Remark 205. *The series*

$$1 - \frac{1}{2} + \frac{1}{3} - \frac{1}{4} + \cdots$$

is one of the most important standard examples of a convergent alternating series. It does converge, but not absolutely. Thus it already points toward a topic that will be important later: the distinction between absolute and conditional convergence.

8.4 Absolute Convergence

For series with alternating signs, it is often not enough to consider only the behavior of the terms themselves. An important stronger form of convergence is therefore *absolute convergence.* It describes series whose absolute values already converge. Such series behave particularly well and can be handled much more easily in many situations than merely conditionally convergent series.

Definition 206 (Absolute convergence). *The series*

$$\sum_{n=0}^{\infty} a_n$$

is called absolutely convergent *if the series of absolute values*

$$\sum_{n=0}^{\infty} |a_n|$$

converges.

In other words, a series is absolutely convergent precisely when

$$\sum_{n=0}^{\infty} |a_n| < \infty.$$

Theorem 207 (Absolute convergence implies convergence). *If the series*

$$\sum_{n=0}^{\infty} |a_n|$$

converges, then the series

$$\sum_{n=0}^{\infty} a_n$$

also converges.

Proof. We use the Cauchy criterion for series.

Assume that

$$\sum_{n=0}^{\infty} |a_n|$$

converges. Then for every $\epsilon > 0$ there exists an $N \in \mathbb{N}$ such that

$$\sum_{k=m}^{n} |a_k| < \epsilon \qquad \text{for all } n \geq m \geq N.$$

By the triangle inequality, it follows for the same indices that

$$\left|\sum_{k=m}^{n} a_k\right| \leq \sum_{k=m}^{n} |a_k| < \epsilon.$$

Thus the series

$$\sum_{n=0}^{\infty} a_n$$

also satisfies the Cauchy criterion for series and therefore converges. □

Remark 208. *Absolute convergence is therefore a stronger property than ordinary convergence. There are* *series that converge but do not converge absolutely, such as the alternating harmonic series*

$$\sum_{n=0}^{\infty} \frac{(-1)^n}{n+1}.$$

Theorem 209 (Comparison test). *Let (b_n) be a sequence of nonnegative real numbers such that*

$$\sum_{n=0}^{\infty} b_n < \infty,$$

and let (a_n) be a sequence of real numbers satisfying

$$|a_n| \leq b_n \qquad \textit{for all } n \in \mathbb{N}$$

Then the series

$$\sum_{n=0}^{\infty} a_n$$

is absolutely convergent.

Proof. From the assumption $|a_n| \leq b_n$, it follows that

$$0 \leq |a_n| \leq b_n \qquad \text{for all } n \in \mathbb{N}.$$

Since the series $\sum_{n=0}^{\infty} b_n$ converges and all its terms are nonnegative, the sequence of its partial sums is bounded by the theorem on monotonically increasing series.

The partial sums of the series $\sum |a_n|$ are also monotonically increasing, because all terms $|a_n|$ are nonnegative. Moreover, for every partial sum we have

$$\sum_{k=0}^{n} |a_k| \leq \sum_{k=0}^{n} b_k.$$

Since the right-hand side is bounded, the left-hand side is bounded as well. Hence the series

$$\sum_{n=0}^{\infty} |a_n|.$$

converges. Therefore, the series $\sum_{n=0}^{\infty} a_n$ is absolutely convergent. □

Remark 210. *The comparison test is one of the most important tools in the theory of series. It makes it possible to decide the convergence of an often complicated series by comparing it with a simpler series whose convergence is already known.*

Example 211 (An example of an absolutely convergent series)**.** *Consider the series*

$$\sum_{n=0}^{\infty} \frac{(-1)^n}{(n+1)^2}.$$

For the absolute values of its terms, we have

$$\left|\frac{(-1)^n}{(n+1)^2}\right| = \frac{1}{(n+1)^2}.$$

Thus the series of absolute values is the series

$$\sum_{n=0}^{\infty} \frac{1}{(n+1)^2},$$

and this series is convergent. Consequently, the original series is absolutely convergent.

This shows that a series may not only converge, but may converge so strongly that even the series of its absolute values is convergent.

A particularly important method for proving absolute convergence is to compare successive terms. If, from some index onward, the terms of a series become smaller in a geometric way, then the whole series can be controlled. This leads to the ratio test.

Theorem 212 (Ratio test). *Let (a_n) be a sequence of real numbers. Suppose that there exist an $n_0 \in \mathbb{N}$ and a number $r \in \mathbb{R}$ with*

$$0 < r < 1,$$

such that

$$\left|\frac{a_{n+1}}{a_n}\right| \leq r \qquad \text{for all } n \geq n_0$$

holds. Then the series

$$\sum_{n=0}^{\infty} a_n$$

is absolutely convergent.

Proof. We first consider the absolute values of the terms. From

$$\left|\frac{a_{n+1}}{a_n}\right| \leq r$$

it follows for all $n \geq n_0$ that

$$|a_{n+1}| \leq r\,|a_n|.$$

Applying this estimate repeatedly, we obtain for $k \in \mathbb{N}$

$$|a_{n_0+k}| \leq r^k\,|a_{n_0}|.$$

Thus the tail of the series of absolute values is dominated by a geometric series:

$$\sum_{k=0}^{\infty} |a_{n_0+k}| \leq |a_{n_0}| \sum_{k=0}^{\infty} r^k.$$

Since $0 < r < 1$, the geometric series $\sum_{k=0}^{\infty} r^k$ converges. Hence the series

$$\sum_{k=0}^{\infty} |a_{n_0+k}|$$

also converges. Therefore the series $\sum_{n=0}^{\infty} |a_n|$ is convergent, and hence

$$\sum_{n=0}^{\infty} a_n$$

is absolutely convergent. □

Remark 213. *The ratio test is particularly useful when the terms involve factorials, powers, or products. In such cases, the quotient can often be computed very easily and compared with a geometric series.*

Example 214 (An example of absolute convergence by the ratio test). *Consider the series*

$$\sum_{n=0}^{\infty} \frac{1}{n!}.$$

For the terms $a_n = \frac{1}{n!}$, *we have*

$$\left|\frac{a_{n+1}}{a_n}\right| = \frac{1/(n+1)!}{1/n!} = \frac{1}{n+1}.$$

Thus, for all $n \geq 1$,

$$\left|\frac{a_{n+1}}{a_n}\right| \leq \frac{1}{2}.$$

Therefore, by the ratio test, the series is absolutely convergent.

Remark 215. *This example is particularly important because the factorial in the denominator grows very quickly, and therefore the terms tend rapidly to* 0.

8.5 The Rearrangement Theorem for Series

For infinite series, the order of the summands plays an important role. If the terms of a series are arranged in a different order, one speaks of a *rearrangement* or *rearrangement of the series.* For finitely many summands, the order is irrelevant because of the commutative law.

For infinite series, however, this is no longer self-evident in general.

Precisely for this reason, it is a deep and very useful result that absolutely convergent series do not lose their convergence behavior under rearrangement: the terms may be reordered arbitrarily without changing the limit.

Definition 216 (Rearrangement of a series). *Let (a_n) be a sequence of real numbers and let $\alpha : \mathbb{N} \to \mathbb{N}$ be a bijection. Then*

$$\sum_{n=0}^{\infty} a_{\alpha(n)}$$

is called a rearrangement *of the series*

$$\sum_{n=0}^{\infty} a_n.$$

Theorem 217 (Rearrangement theorem for series). *Let*

$$\sum_{n=0}^{\infty} a_n$$

be an absolutely convergent series. Then every rearrangement of this series also converges absolutely, and it converges to the same limit.

Proof. Let

$$S := \sum_{n=0}^{\infty} a_n$$

and let $\alpha : \mathbb{N} \to \mathbb{N}$ be a bijection. We define

$$s_m := \sum_{k=0}^{m} a_{\alpha(k)}.$$

We have to show that $s_m \to S$ as $m \to \infty$.

Since the series $\sum_{n=0}^{\infty} |a_n|$ converges, the Cauchy criterion implies that for every $\epsilon > 0$ there exists an $n_0 \in \mathbb{N}$ such that

$$\sum_{k=n_0}^{\infty} |a_k| < \epsilon.$$

In particular, the remainder of the original series is then small as well:

$$\left| S - \sum_{k=0}^{n_0-1} a_k \right| = \left| \sum_{k=n_0}^{\infty} a_k \right| \leq \sum_{k=n_0}^{\infty} |a_k| < \epsilon.$$

Since α is a bijection, the finitely many indices

$$0, 1, \ldots, n_0 - 1$$

all occur eventually in the sequence $\alpha(0), \alpha(1), \alpha(2), \ldots$. Hence there exists an $N \in \mathbb{N}$ such that

$$\{0, 1, \ldots, n_0 - 1\} \subset \{\alpha(0), \alpha(1), \ldots, \alpha(N)\}.$$

For every $m \geq N$, the partial sum s_m therefore already contains all first n_0 terms of the original series. The difference from S then consists only of terms with index $\geq n_0$. Thus

$$|s_m - S| \leq \sum_{k=n_0}^{\infty} |a_k| < \epsilon.$$

It follows that

$$\lim_{m\to\infty} s_m = S.$$

Thus the rearranged series converges to the same limit as the original series.

Since the series is absolutely convergent, the same statement also applies to the series of absolute values $\sum_{n=0}^{\infty} |a_n|$. Every rearrangement of this series also converges to the same value. Consequently, the rearranged series $\sum_{n=0}^{\infty} a_{\alpha(n)}$ is absolutely convergent as well. □

Remark 218. *The rearrangement theorem for series shows a fundamental difference between absolutely and only conditionally convergent series: for absolutely convergent series, the order of the terms is unproblematic. This is of great importance later in many passages to limits.*

8.6 Rearrangements and Conditional Convergence

The theorem on rearrangements of series relies essentially on absolute convergence. If absolute convergence is absent, then the order of the summands is no longer harmless: a rearrangement may change the limit or even destroy convergence altogether.

A classical example is the alternating harmonic series

$$\sum_{n=0}^{\infty} \frac{(-1)^n}{n+1} = 1 - \frac{1}{2} + \frac{1}{3} - \frac{1}{4} + \frac{1}{5} - \frac{1}{6} + \cdots .$$

This series converges, but it does not converge absolutely. For precisely this reason, its terms can be rearranged in such a way that a different limit is obtained.

Example 219 (A rearrangement can change the limit). *Consider the rearrangement*

$$1+\frac{1}{3}-\frac{1}{2}+\frac{1}{5}+\frac{1}{7}-\frac{1}{4}+\frac{1}{9}+\frac{1}{11}-\frac{1}{6}+\cdots.$$

Here the same terms as in the alternating harmonic series are used, but in a different order: two positive terms are taken first, followed by one negative term.

Thus this new series is a genuine rearrangement of the original one. In general, however, it does not have the same limit as the original alternating harmonic series. This example therefore shows that the rearrangement theorem for series no longer applies to series that are merely convergent but not absolutely convergent.

Remark 220. *In fact, this behavior is typical of conditionally convergent series: by rearranging the terms, one can change their limit, and with somewhat more effort one can even force any prescribed real number to be the limit. Absolute convergence is therefore precisely the condition that makes rearrangements unproblematic.*

Example 221 (A rearrangement with the same limit). *Consider the absolutely convergent series*

$$\sum_{n=0}^{\infty}\frac{(-1)^n}{2^n}=1-\frac{1}{2}+\frac{1}{4}-\frac{1}{8}+\cdots.$$

By the rearrangement theorem for series, its terms may be rearranged arbitrarily without changing the limit.

For example, the rearranged series

$$1+\frac{1}{4}-\frac{1}{2}+\frac{1}{16}-\frac{1}{8}+\frac{1}{64}-\frac{1}{32}+\cdots$$

is also absolutely convergent and has the same limit as the original series.

The original series is a geometric series with ratio $q=-\frac{1}{2}$. *Hence*

$$\sum_{n=0}^{\infty}\frac{(-1)^n}{2^n}=\frac{1}{1-(-\frac{1}{2})}=\frac{2}{3}.$$

Consequently, every rearrangement of this series also converges to $\frac{2}{3}$.

Remark 222. *This example illustrates the contrast with the conditionally convergent case: for absolutely convergent series, the limit is preserved under every rearrangement.*

8.7 Exercises and Solutions

Exercise 223. *Investigate the series*

$$\sum_{n=1}^{\infty}\frac{(-1)^{n+1}}{n}$$

for convergence.

Solution 224. *We write the series in the form*

$$\sum_{n=1}^{\infty}(-1)^{n+1}a_n \qquad \text{where} \qquad a_n = \frac{1}{n}.$$

Then (a_n) is a monotonically decreasing sequence of nonnegative numbers and

$$\lim_{n\to\infty} a_n = 0.$$

Therefore, by the Leibniz convergence criterion, the series

$$\sum_{n=1}^{\infty}\frac{(-1)^{n+1}}{n}$$

converges.

However, it does not converge absolutely, since the series of absolute values is

$$\sum_{n=1}^{\infty}\left|\frac{(-1)^{n+1}}{n}\right| = \sum_{n=1}^{\infty}\frac{1}{n}.$$

This is the harmonic series, and it diverges. Hence the given series is only conditionally convergent.

Exercise 225. *Investigate the series*

$$\sum_{n=0}^{\infty}\frac{1}{2^n}$$

for convergence.

Solution 226. *This is a geometric series with ratio*

$$q = \frac{1}{2}.$$

Since

$$|q| < 1,$$

the series converges by the theorem on geometric series. Using the summation formula, we also obtain

$$\sum_{n=0}^{\infty} \frac{1}{2^n} = \frac{1}{1 - \frac{1}{2}} = 2.$$

Thus the series is convergent and has sum 2*.*

Exercise 227. *Use the ratio test to show that the series*

$$\sum_{n=0}^{\infty} \frac{1}{n!}$$

converges absolutely.

Solution 228. *We consider the sequence*

$$a_n = \frac{1}{n!}.$$

Then, for all $n \in \mathbb{N}$*, we have*

$$\left| \frac{a_{n+1}}{a_n} \right| = \frac{1/(n+1)!}{1/n!} = \frac{1}{n+1}.$$

In particular, for all $n \geq 1$*,*

$$\left| \frac{a_{n+1}}{a_n} \right| \leq \frac{1}{2}.$$

Thus the hypothesis of the ratio test is satisfied. Hence the series

$$\sum_{n=0}^{\infty} \frac{1}{n!}$$

converges absolutely.

Chapter 9

Point Sets

So far, we have studied the real numbers as a number system and have become familiar with the most important properties of their order and completeness. This lays the foundation for no longer considering only individual numbers or sequences, but entire sets of points on the number line in a systematic way.

Such point sets occur constantly in analysis. Even simple questions about the behavior of sequences, about limits, or about the existence of extrema repeatedly lead back to sets of real numbers. Point sets will become especially important in the next chapter, when we study functions and continuity: there, neighborhoods, intervals, and the relative position of points play a central role. Before moving on to functions, we therefore first clarify the basic concepts concerning subsets of $\mathbb{R}$.

9.1 Intervals

The simplest and at the same time most important point sets in analysis are intervals. They describe connected parts of the real number line and form the natural language in which many analytical statements are formulated.

Definition 229. *Let $a, b \in \mathbb{R}$ with $a < b$. Then*

$$(a, b) := \{x \in \mathbb{R} \mid a < x < b\}$$

is called the open interval *from a to b.*

Correspondingly,

$$[a, b] := \{x \in \mathbb{R} \mid a \leq x \leq b\},$$

$$(a, b] := \{x \in \mathbb{R} \mid a < x \leq b\},$$
$$[a, b) := \{x \in \mathbb{R} \mid a \leq x < b\}$$

are called the closed *and* half-open intervals, *respectively.*

Open intervals describe regions without boundary points, closed intervals contain their boundary points, and half-open intervals lie between these two cases. This distinction is not merely a formal subtlety, but will later be decisive again and again in the study of limits, continuity, and integrals.

In addition to these finite intervals, we also consider unbounded intervals, such as

$$(a, \infty), \quad [a, \infty), \quad (-\infty, b), \quad (-\infty, b],$$

as well as the entire real number line $\mathbb{R}$ itself. Such *improper intervals* are particularly useful when describing unbounded sets or behavior "toward infinity."

In the following, we shall use the various types of intervals as basic building blocks for point sets. They provide the first important class of examples and will also help us later to classify concepts such as openness, closedness, and boundary points in a precise way.

9.2 Supremum and Infimum of Point Sets

In this section, we consider arbitrary subsets of the real numbers and ask how their position on the number line can be described as precisely as possible. Of particular importance are the least upper bound and the greatest lower bound of a set. These concepts generalize the notions of maximum and minimum and remain meaningful even when a largest or smallest element is not contained in the set.

Definition 230 (Upper and lower bounds). *Let $M \subseteq \mathbb{R}$ and $s \in \mathbb{R}$.*

Then s is called an upper bound *of M if, for all $x \in M$,*

$$x \leq s.$$

Analogously, $t \in \mathbb{R}$ is called a lower bound *of M if, for all $x \in M$,*

$$t \leq x.$$

Definition 231 (Boundedness). *A set $M \subseteq \mathbb{R}$ is called* bounded above *if it has an upper bound.*

It is called bounded below *if it has a lower bound.*

A set is called bounded *if it is both bounded above and bounded below.*

Definition 232 (Supremum). *Let $M \subseteq \mathbb{R}$ be nonempty and bounded above. A number $s \in \mathbb{R}$ is called the* supremum *of M if the following conditions hold:*

1. *s is an upper bound of M,*

2. *for every upper bound u of M, one has $s \leq u$.*

The supremum of M is denoted by $\sup M$.

Definition 233 (Infimum). *Let $M \subseteq \mathbb{R}$ be nonempty and bounded below. A number $i \in \mathbb{R}$ is called the* infimum *of M if the following conditions hold:*

1. *i is a lower bound of M,*

2. *for every lower bound l of M, one has $l \leq i$.*

The infimum of M is denoted by $\inf M$.

Intuitively, $\sup M$ is therefore the smallest number that still lies above the entire set, while $\inf M$ is the largest number that still lies below the entire set.

Theorem 234 (Completeness of the real numbers). *Every nonempty subset $M \subseteq \mathbb{R}$ that is bounded above has a supremum. Every nonempty subset $M \subseteq \mathbb{R}$ that is bounded below has an infimum.*

Proof. We first prove the statement about the supremum.

Let $M \subseteq \mathbb{R}$ be nonempty and bounded above. Then there exists at least one upper bound of M. We denote the set of all upper bounds of M by

$$U := \{u \in \mathbb{R} \mid u \text{ is an upper bound of } M\}$$

.

Since M is bounded above, the set U is nonempty. Moreover, M itself is nonempty. Choose an arbitrary $x_0 \in M$ and an arbitrary $u_0 \in U$. Then for every $x \in M$ we have the estimate $x \leq u_0$, and in particular $x_0 \leq u_0$. Thus x_0 is a lower bound of U, because every upper bound of M is greater than or equal to every element of M.

Hence U is nonempty and bounded below. By the completeness axiom, U has an infimum. We set

$$s := \inf U.$$

We claim that $s = \sup M$.

First we show that s is an upper bound of M. Suppose this were not the case. Then there would exist an $x \in M$ such that

$$x > s.$$

Then every number u with $u \in U$, being an upper bound of M, must necessarily be at least as large as x, so $u \geq x > s$. Thus x would be a lower bound of U that is greater than s. This contradicts the property of s as the greatest lower bound of U.

Therefore s is indeed an upper bound of M.

Now let u be any upper bound of M. Then, by definition, $u \in U$. Since $s = \inf U$, it follows that

$$s \leq u.$$

Thus s is the least upper bound of M, hence

$$s = \sup M.$$

The statement about the infimum is obtained analogously, or directly by applying what has just been proved to the set

$$-M := \{-x \mid x \in M\}.$$

If M is nonempty and bounded below, then $-M$ is nonempty and bounded above. Therefore $-M$ has a supremum. For $i := -\sup(-M)$, the number i is then the infimum of M. □

Remark 235. *The supremum of a set need not belong to the set itself. For example, for*

$$M = (0, 1)$$

we have

$$\sup M = 1 \qquad \textit{and} \qquad \inf M = 0,$$

although neither 0 *nor* 1 *is an element of* M.

If the supremum of a set belongs to the set itself, then it is also its maximum. Correspondingly, the infimum of a set is its minimum precisely when it is contained in the set.

9.3 Limit Superior and Limit Inferior

For sequences, ordinary limits do not always appear immediately. Often the terms of a sequence oscillate or approach a value only from one side. In such situations, the *limit superior* and the *limit inferior* are particularly useful concepts. They describe, in a sense, the uppermost and lowermost cluster regions of a sequence and will reappear later in the study of functions and continuity.

Definition 236 (Tails of a sequence). *Let $(a_n)_{n\in\mathbb{N}}$ be a real sequence. For each $n \in \mathbb{N}$, we consider the tail set*

$$A_n := \{a_k \mid k \geq n\}.$$

Then

$$s_n := \sup A_n \quad \textit{and} \quad i_n := \inf A_n$$

is called the upper *respectively* lower *bound of the n-th tail of the sequence.*

Definition 237 (Limit superior and limit inferior). *If (a_n) is a sequence bounded above, then*

$$\limsup_{n\to\infty} a_n := \lim_{n\to\infty} \sup_{k\geq n} a_k$$

is called the limit superior *of the sequence.*

If (a_n) is bounded below, then

$$\liminf_{n\to\infty} a_n := \lim_{n\to\infty} \inf_{k\geq n} a_k$$

is called the limit inferior *of the sequence.*

The sequences (s_n) and (i_n) are monotone:

$$s_{n+1} \leq s_n, \qquad i_{n+1} \geq i_n.$$

Therefore, provided they are bounded, each of them has a limit. This limit is precisely the supremum respectively infimum of the possible cluster regions of the sequence.

Theorem 238 (Characterization of the limit superior). *For a real sequence (a_n) and a number $a \in \mathbb{R}$, we have*

$$\limsup_{n\to\infty} a_n = a$$

if and only if, for every $\varepsilon > 0$, the following two conditions are satisfied:

1. *there are only finitely many $n \in \mathbb{N}$ such that*

$$a_n \geq a + \varepsilon,$$

that is,

$$a_n < a + \varepsilon \quad \textit{for all but finitely many } n,$$

2. *there are infinitely many $m \in \mathbb{N}$ such that*

$$a_m > a - \varepsilon.$$

Proof. We set

$$s_n := \sup_{k\geq n} a_k.$$

Then the assertion ($\limsup a_n = a$) is precisely the statement that $s_n \to a$.

Necessity. Assume that $s_n \to a$. Let $\varepsilon > 0$.

Since $s_n \to a$, there exists an N such that

$$s_n < a + \varepsilon \qquad \text{for all } n \geq N.$$

But for every $n \geq N$ we have $a_n \leq s_n$, hence

$$a_n < a + \varepsilon \qquad \text{for all } n \geq N.$$

This proves the first condition.

Moreover, since $s_n \to a$, there also exists an N' such that

$$s_n > a - \varepsilon \qquad \text{for all } n \geq N'.$$

Since s_n is the supremum of the tail set $\{a_k \mid k \geq n\}$, for every $n \geq N'$ there exists an index $m \geq n$ such that

$$a_m > a - \varepsilon.$$

Thus there cannot be only finitely many such indices; there are infinitely many m with $a_m > a - \varepsilon$.

Sufficiency. Conversely, suppose that the two conditions hold for every $\varepsilon > 0$.

We show that $s_n \to a$. To this end, let $\varepsilon > 0$.

By the first condition, there exists an N such that

$$a_k < a + \varepsilon \qquad \text{for all } k \geq N.$$

Then, for every $n \geq N$, the entire tail set $\{a_k \mid k \geq n\}$ is bounded above by $a + \varepsilon$, and therefore

$$s_n \leq a + \varepsilon \qquad \text{for all } n \geq N.$$

The second condition implies that for every $n \in \mathbb{N}$ there exists an $m \geq n$ such that

$$a_m > a - \varepsilon.$$

Thus every tail set contains an element greater than $a - \varepsilon$, and hence

$$s_n \geq a - \varepsilon \qquad \text{for all } n \in \mathbb{N}.$$

Together, for all $n \geq N$ we obtain

$$a - \varepsilon \leq s_n \leq a + \varepsilon.$$

Since $\varepsilon > 0$ was arbitrary, it follows that $s_n \to a$, and therefore

$$\limsup_{n\to\infty} a_n = a.$$

□

Theorem 239 (Characterization of the limit inferior). *For a real sequence (a_n) and a number $a \in \mathbb{R}$, one has*

$$\liminf_{n\to\infty} a_n = a$$

if and only if, for every $\varepsilon > 0$, the following two conditions are satisfied:

1. *there are only finitely many* $n \in \mathbb{N}$ *such that*

$$a_n \leq a - \varepsilon,$$

that is,

$$a_n > a - \varepsilon \quad \text{for all but finitely many } n,$$

2. *there are infinitely many* $m \in \mathbb{N}$ *such that*

$$a_m < a + \varepsilon.$$

Proof. The proof is entirely analogous to the proof for the limit superior. Consider

$$i_n := \inf_{k \geq n} a_k$$

and show that $i_n \to a$ holds if and only if the two ε-conditions above are satisfied. □

Remark 240. *If, in fact,*

$$\limsup_{n \to \infty} a_n = \liminf_{n \to \infty} a_n = a,$$

then the sequence has the ordinary limit

$$\lim_{n \to \infty} a_n = a.$$

Thus the limit superior and the limit inferior refine the notion of a limit: they also capture the behavior of sequences that do not have a limit.

Chapter 10

Continuous Functions

In the previous chapters, we became familiar with the real numbers as an ordered and complete field, studied limits of sequences, and developed some first fundamental tools of analysis. We now turn to a new central object: functions.

Functions describe dependencies. They assign to each admissible input value a uniquely determined output value and thereby form the language in which almost all further concepts of analysis are formulated. In particular, we shall later study limits and continuity of functions. The underlying idea is simple and at the same time profound: a function should behave as regularly as possible locally, so that small changes in the input produce only small changes in its value.

Before introducing the notion of a limit for functions, however, we need a precise definition of what we mean by a function. This is exactly where we begin in this chapter.

10.1 Functions

Definition 241. *Let $D \subset \mathbb{R}$. A real-valued* function *on D is a mapping*

$$f : D \to \mathbb{R}.$$

The set D is called the domain *of f. The* graph *of f is the set of points*

$$G_f := \{(x, y) \in D \times \mathbb{R} : y = f(x)\}.$$

In other words: to each $x \in D$ there corresponds exactly one value $f(x) \in \mathbb{R}$. The graph of a function is the set of all points whose second coordinate is determined by the rule defining the function.

Example 242 (The constant function)**.** *Let $D = \mathbb{R}$ and*

$$f(x) = 2 \qquad \text{for all } x \in \mathbb{R}.$$

Then f is a constant function. Its graph is a horizontal line.

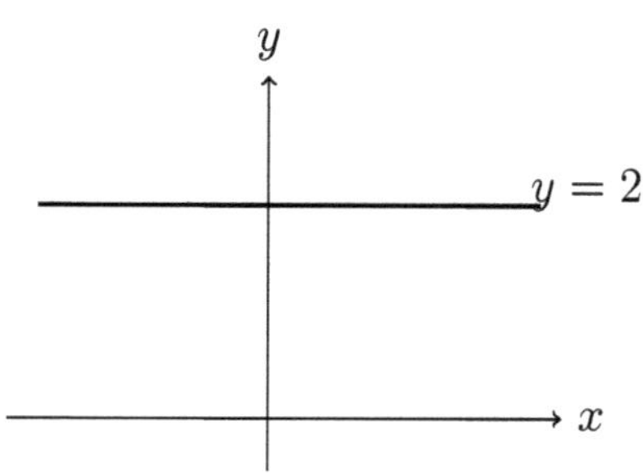

Example 243 (The identity function). *Let $D = \mathbb{R}$ and*

$$g(x) = x \qquad \text{for all } x \in \mathbb{R}.$$

This function is called the identity function. *Its graph is the diagonal line through the origin.*

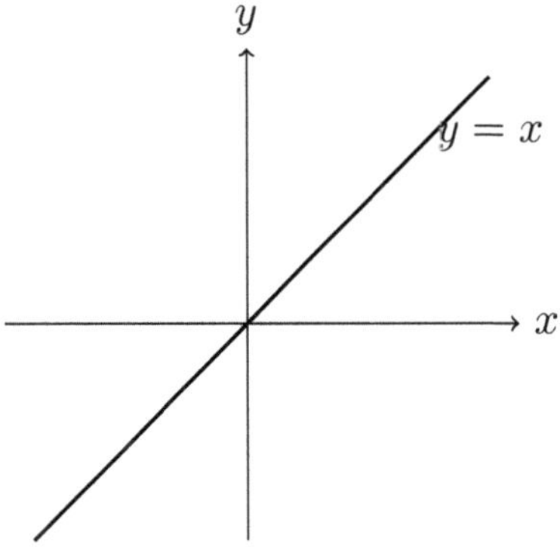

Example 244 (The quadratic function). *Let $D = \mathbb{R}$ and*

$$h(x) = x^2 \qquad \text{for all } x \in \mathbb{R}.$$

This function is a first important example of a nonlinear function. Its graph is a parabola opening upward.

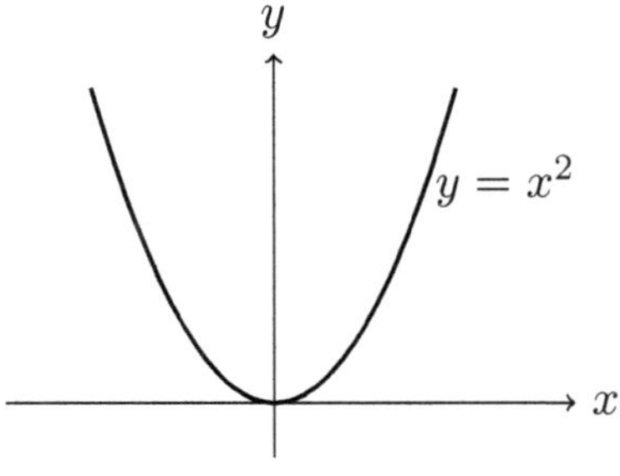

Example 245 (The absolute value function). *Let $D = \mathbb{R}$ and*

$$a(x) = |x| \qquad \text{for all } x \in \mathbb{R}.$$

The absolute value function is an important example because it has a corner at $x = 0$. Its graph is V-shaped.

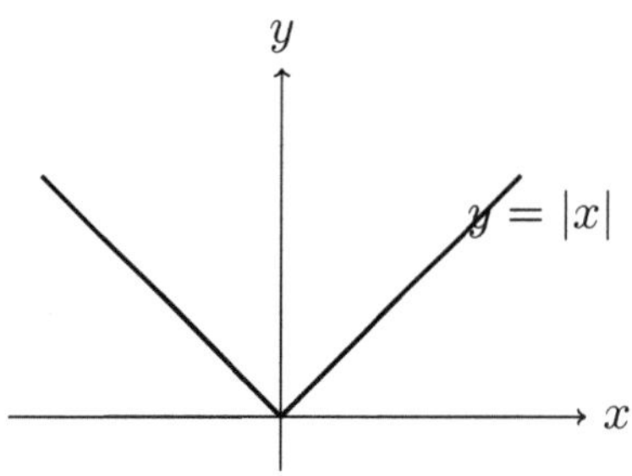

These first examples already show that functions can have very different shapes: constant, linear, quadratic, or with a corner. In the remainder of this chapter, we will learn how to combine functions, investigate limits of functions, and finally define continuity precisely.

10.1.1 Operations on Functions

Let $f, g\colon D \to \mathbb{R}$ be two real-valued functions with the same domain D. Then we can combine them pointwise.

Definition 246. *Let $f, g\colon D \to \mathbb{R}$ be real-valued functions on D, and let $\lambda \in \mathbb{R}$. We define the* sum, *the* scalar multiple, *and the* product *of f and g by*

$$f + g\colon D \to \mathbb{R}, \qquad (f+g)(x) := f(x) + g(x),$$

$$\lambda f\colon D \to \mathbb{R}, \qquad (\lambda f)(x) := \lambda\, f(x),$$
$$fg\colon D \to \mathbb{R}, \qquad (fg)(x) := f(x)\, g(x)$$

for all $x \in D$.

Moreover, if

$$E := \{x \in D : g(x) \neq 0\},$$

then we define the quotient *of f and g as the function*

$$\frac{f}{g}\colon E \to \mathbb{R}, \qquad \left(\frac{f}{g}\right)(x) := \frac{f(x)}{g(x)}$$

for all $x \in E$.

These definitions are examples of so-called *pointwise* operations: the respective arithmetic operation is applied not to the functions as whole objects, but to their function values.
A particularly important case is the repeated application of these operations to the identity function

$$\mathrm{id}_{\mathbb{R}}\colon \mathbb{R} \to \mathbb{R}, \qquad \mathrm{id}_{\mathbb{R}}(x) = x,$$

and to the constant function

$$\mathbf{1}\colon \mathbb{R} \to \mathbb{R}, \qquad \mathbf{1}(x) = 1.$$

Starting from these two functions, all *rational functions* in one variable can be generated by addition, subtraction, multiplication, and forming quotients; that is, all functions that can be written as the quotient of two polynomials.

10.1.2 Composition of Functions

Another important way to combine functions is their *composition.*

Definition 247. *Let*

$$f\colon D \to \mathbb{R} \qquad \text{and} \qquad g\colon E \to \mathbb{R}$$

be functions with $f(D) \subset E$. Then the function defined by

$$g \circ f\colon D \to \mathbb{R}, \qquad (g \circ f)(x) := g(f(x))$$

for all $x \in D$ is called the composition *of g with f.*

Composition therefore means the following: first, the value x is processed by the function f, and the resulting value is then used as the input for g.

Example 248. *Let*

$$f\colon \mathbb{R} \to \mathbb{R}, \qquad f(x) = x + 1,$$

and

$$g\colon \mathbb{R} \to \mathbb{R}, \qquad g(x) = x^2.$$

Then

$$(g \circ f)(x) = g(f(x)) = g(x + 1) = (x + 1)^2.$$

Thus the composition $g \circ f$ is the function

$$x \mapsto (x + 1)^2.$$

10.2 Limits of Functions

After studying sequences and their limits, we now transfer the same idea to functions. For sequences, one observes what happens to the terms of a numerical sequence as the indices become large. For functions, by contrast, one asks how the function values behave when the input values approach a certain point. The notion of a limit for functions is therefore the natural extension of the notion of a limit for sequences.

A central role is played by the question of whether a point can be approached at all by values from the domain under consideration. For this purpose, we first introduce the notion of an accumulation point.

Definition 249. *Let $D \subset \mathbb{R}$. A point $a \in \mathbb{R}$ is called an* accumulation point *of D if there exists a sequence (a_n) of elements of D such that*

$$\lim_{n\to\infty} a_n = a.$$

Thus every point $a \in D$ is an accumulation point of D, since one may choose, for instance, the constant sequence $a_n = a$ for all $n \in \mathbb{N}$. However, points outside of D may also be accumulation points. For example, 0 is an accumulation point of

$$D = (0, 1),$$

although $0 \notin D$, because the sequence $a_n = \frac{1}{n}$ lies in D and satisfies $a_n \to 0$.

Definition 250. *Let $f\colon D \to \mathbb{R}$ be a real-valued function on $D \subset \mathbb{R}$, and let $a \in \mathbb{R}$ be an accumulation point of D. We write*

$$\lim_{x \to a} f(x) = c,$$

if for every sequence (x_n) with $x_n \in D$ and

$$\lim_{n \to \infty} x_n = a$$

we also have

$$\lim_{n \to \infty} f(x_n) = c.$$

Thus the limit of the function is equal to c precisely when all sequences from the domain that approach a are mapped by f to the same limit c. This formulation is particularly useful because it connects directly to the concept of limits of sequences.

Definition 251. *Let $f\colon D \to \mathbb{R}$ be a real-valued function on $D \subset \mathbb{R}$, and let $a \in \mathbb{R}$ be an accumulation point of D. We write*

$$\lim_{x \downarrow a} f(x) = c,$$

if for every sequence (x_n) with $x_n \in D$, $x_n > a$ for all n, and

$$\lim_{n \to \infty} x_n = a$$

also

$$\lim_{n \to \infty} f(x_n) = c$$

holds.

Analogously, we write

$$\lim_{x \uparrow a} f(x) = c,$$

if for every sequence (x_n) *with* $x_n \in D$, $x_n < a$ *for all* n, *and*

$$\lim_{n\to\infty} x_n = a$$

also

$$\lim_{n\to\infty} f(x_n) = c$$

holds.

The right-hand and left-hand limits therefore describe the behavior of the function near a when one approaches the point from only one side.

Definition 252. *Let* $f\colon D \to \mathbb{R}$ *be a real-valued function on* $D \subset \mathbb{R}$. *We write*

$$\lim_{x\to\infty} f(x) = c,$$

if D *is unbounded above and, for every sequence* (x_n) *with* $x_n \in D$ *and*

$$\lim_{n\to\infty} x_n = \infty$$

also

$$\lim_{n\to\infty} f(x_n) = c$$

holds.

Analogously, we write

$$\lim_{x\to-\infty} f(x) = c,$$

if D is unbounded below and, for every sequence (x_n) with $x_n \in D$ and

$$\lim_{n\to\infty} x_n = -\infty$$

also

$$\lim_{n\to\infty} f(x_n) = c$$

holds.

This also precisely captures the behavior of a function for very large or very small arguments.

Example 253 (A polynomial). *For*

$$f(x) = x^2$$

we have

$$\lim_{x\to 2} x^2 = 4.$$

Indeed, if $x_n \to 2$, then the continuity of the squaring function implies that $x_n^2 \to 4$.

Moreover,

$$\lim_{x\to\infty} x^2 = \infty.$$

Example 254 (A rational function). *For*

$$f(x) = \frac{x^2 - 1}{x - 1} \qquad (x \neq 1)$$

we have

$$\frac{x^2 - 1}{x - 1} = \frac{(x-1)(x+1)}{x-1} = x + 1$$

for all $x \neq 1$*. Hence*

$$\lim_{x \to 1} \frac{x^2 - 1}{x - 1} = 2.$$

This is a typical example showing that a function need not have a value at a point in order to have a limit there.

Example 255 (One-sided limits). *For the absolute value function*

$$f(x) = |x|$$

at $a = 0$ *we have both*

$$\lim_{x \downarrow 0} |x| = 0 \qquad \text{and} \qquad \lim_{x \uparrow 0} |x| = 0.$$

Therefore,

$$\lim_{x \to 0} |x| = 0.$$

Example 256 (Different left- and right-hand limits). *Consider the function*

$$f(x) = \begin{cases} 0, & x < 0, \\ 1, & x \geq 0. \end{cases}$$

Then

$$\lim_{x \uparrow 0} f(x) = 0 \qquad \text{and} \qquad \lim_{x \downarrow 0} f(x) = 1.$$

Thus the two-sided limit $\lim_{x \to 0} f(x)$ *does not exist.*

These examples already illustrate the basic idea: limits of functions do not describe the exact value of the function at a point, but rather the behavior of the function in a neighborhood of that point. This is precisely the foundation on which the concept of continuity is built in the next step.

10.3 Continuity of a Function

The concept of a limit leads directly to one of the most important fundamental notions in analysis: continuity. A function is called continuous if its value at a point agrees exactly with the limit approached by the function at that point. Intuitively, this means that near a point the function behaves without jumps, gaps, or sudden deviations.

Definition 257. *Let $f\colon D \to \mathbb{R}$ be a function, and let $a \in D$. Then f is called* continuous at a *if*

$$\lim_{x \to a} f(x) = f(a).$$

The function f is called continuous on D *if it is continuous at every point of D.*

Continuity is therefore a local property: it is checked point by point. Only when it holds everywhere in the domain do we speak of a function that is continuous on all of D.

Example 258 (A constant function). *The function*

$$f\colon \mathbb{R} \to \mathbb{R}, \qquad f(x) = 2,$$

is continuous at every point of $\mathbb{R}$, *since for every sequence* (x_n) *with* $x_n \to a$ *we immediately have*

$$f(x_n) = 2 \quad \textit{for all } n,$$

and hence also

$$\lim_{n\to\infty} f(x_n) = 2 = f(a).$$

Example 259 (The square function). *The function*

$$g\colon \mathbb{R} \to \mathbb{R}, \qquad g(x) = x^2,$$

is also continuous on $\mathbb{R}$. *Indeed, for every sequence* (x_n) *with* $x_n \to a$, *the rules for limits of sequences imply that*

$$x_n^2 \to a^2.$$

Thus

$$\lim_{x\to a} x^2 = a^2 = g(a).$$

The great importance of continuity lies in the fact that it is compatible with the basic operations of analysis. Sums, products, scalar multiples, quotients, and compositions of continuous functions are again continuous. This is precisely what the following theorem shows.

Theorem 260. *Let $f, g\colon D \to \mathbb{R}$ be functions that are continuous at a point $a \in D$, and let $\lambda \in \mathbb{R}$. Then the functions*

$$f+g\colon D \to \mathbb{R}, \qquad \lambda f\colon D \to \mathbb{R}, \qquad fg\colon D \to \mathbb{R}$$

are also continuous at a. In addition, the quotient function

$$\frac{f}{g}\colon E \to \mathbb{R}, \qquad E := \{x \in D : g(x) \neq 0\},$$

is continuous at a, provided that $g(a) \neq 0$.

Proof. We use the sequential criterion for continuity. Let (x_n) be a sequence with $x_n \in D$ and $x_n \to a$. Since f and g are continuous at a, we have

$$f(x_n) \to f(a) \qquad \text{and} \qquad g(x_n) \to g(a).$$

By the arithmetic rules for limits of sequences, it follows that

$$(f+g)(x_n) = f(x_n) + g(x_n) \to f(a) + g(a) = (f+g)(a),$$

$$(\lambda f)(x_n) = \lambda f(x_n) \to \lambda f(a) = (\lambda f)(a),$$
$$(fg)(x_n) = f(x_n)g(x_n) \to f(a)g(a) = (fg)(a).$$

Thus $f+g$, λf, and fg are continuous at a.

For the quotient, assume in addition that $g(a) \neq 0$. Since $g(x_n) \to g(a)$, it follows in particular that $g(x_n) \neq 0$ for all sufficiently large n. Hence the quotient along the sequence is well-defined from some index onward. The arithmetic rules for limits then give

$$\frac{f(x_n)}{g(x_n)} \to \frac{f(a)}{g(a)}.$$

Thus $\frac{f}{g}$ is also continuous at a. □

Theorem 261. *Let*

$$f\colon D \to \mathbb{R} \qquad \textit{and} \qquad g\colon E \to \mathbb{R}$$

be functions with $f(D) \subset E$. Moreover, let $a \in D$ and assume that f is continuous at a and that g is continuous at $f(a)$. Then the composition

$$g \circ f\colon D \to \mathbb{R}$$

is also continuous at a.

Proof. Again, we use the sequential criterion for continuity. Let (x_n) be a sequence with $x_n \in D$ and

$$\lim_{n\to\infty} x_n = a.$$

Since f is continuous at a, it follows that

$$\lim_{n\to\infty} f(x_n) = f(a).$$

Since g is continuous at $f(a)$, we may now apply g to this sequence and obtain

$$\lim_{n\to\infty} g(f(x_n)) = g(f(a)).$$

Hence

$$\lim_{n\to\infty}(g\circ f)(x_n) = (g\circ f)(a).$$

Therefore $g\circ f$ is continuous at a. □

Example 262. *Consider the functions*

$$f\colon \mathbb{R}\to\mathbb{R}, \qquad f(x) = x^2,$$

and

$$g\colon [0,\infty)\to\mathbb{R}, \qquad g(x) = \sqrt{x}.$$

Both functions are continuous on their respective domains. Moreover,

$$f(\mathbb{R}) = [0,\infty),$$

so the composition $g\circ f$ is well-defined.

For every $x\in\mathbb{R}$, we obtain

$$(g\circ f)(x) = g(f(x)) = \sqrt{x^2} = |x|.$$

Thus

$$g\circ f\colon \mathbb{R}\to\mathbb{R}, \qquad x\mapsto |x|.$$

By the preceding theorem, the composition $g\circ f$ is continuous. Hence the continuity of the absolute value function follows as a consequence of the continuity of the square root function and the squaring function.

This example shows how new continuous functions can be constructed from functions that are already known to be continuous. In fact, many important functions in analysis are built in precisely this way.

We have therefore shown that continuity is preserved under the basic algebraic operations. In the next step, we will use this fact to systematically build a large class of continuous functions.

10.4 Exercises and Solutions

Exercise 263. *Using the calculation rules proved in this chapter, show that the polynomial function*

$$p(x) = x^3 - 2x + 5$$

is continuous on all of $\mathbb{R}$.

Solution 264. *The function*

$$p(x) = x^3 - 2x + 5$$

is a sum and difference of polynomial functions. More precisely,

$$p(x) = x^3 + (-2)x + 5.$$

The functions $x \mapsto x$, $x \mapsto x^3$, $x \mapsto (-2)x$, *and the constant function* $x \mapsto 5$ *are continuous on all of* $\mathbb{R}$. *Since sums and scalar multiples of continuous functions are again continuous, it follows that* p *is also continuous on all of* $\mathbb{R}$.

Exercise 265. *Let*

$$f\colon \mathbb{R} \to \mathbb{R}, \qquad f(x) = x^2,$$

and

$$g\colon [0,\infty) \to \mathbb{R}, \qquad g(x) = \sqrt{x}.$$

Compute the composition $g \circ f$ and show that the resulting function is continuous.

Solution 266. *For all $x \in \mathbb{R}$, we have*

$$(g \circ f)(x) = g(f(x)) = g(x^2) = \sqrt{x^2} = |x|.$$

Thus

$$g \circ f\colon \mathbb{R} \to \mathbb{R}, \qquad x \mapsto |x|.$$

Since $f(x) = x^2$ is continuous on $\mathbb{R}$ and $g(x) = \sqrt{x}$ is continuous on $[0,\infty)$, the theorem on the continuity of compositions implies that $g \circ f$ is continuous as well. Hence the absolute value function is continuous on all of $\mathbb{R}$.

Exercise 267. *Consider the function*

$$h\colon \mathbb{R} \to \mathbb{R}, \qquad h(x) = \begin{cases} \dfrac{x^2-1}{x-1}, & x \neq 1, \\ 2, & x = 1. \end{cases}$$

Determine whether h is continuous at $x = 1$.

Solution 268. *For $x \neq 1$, we have*

$$h(x) = \frac{x^2-1}{x-1} = \frac{(x-1)(x+1)}{x-1} = x + 1.$$

Therefore,

$$\lim_{x\to 1} h(x) = \lim_{x\to 1}(x+1) = 2.$$

On the other hand,

$$h(1) = 2.$$

Thus

$$\lim_{x \to 1} h(x) = h(1).$$

Hence h is continuous at $x = 1$.

10.5 The Intermediate Value Theorem

The intermediate value theorem is one of the first results that turns continuity into a genuinely powerful statement. Intuitively, it says that a continuous function cannot simply skip a value on an interval. If its values at the two endpoints lie on different sides of an intermediate value, then the function must attain that value somewhere in between.

This is exactly where the importance of the completeness of the real numbers becomes visible. The theorem holds for functions on real intervals, but not in every ordered number system.

Theorem 269 (Intermediate value theorem). *Let $f\colon [a, b] \to \mathbb{R}$ be a continuous function, and let $y \in \mathbb{R}$ be a value between $f(a)$ and $f(b)$. That is, suppose that either*

$$f(a) \leq y \leq f(b) \qquad \textit{or} \qquad f(b) \leq y \leq f(a).$$

Then there exists a point $c \in [a, b]$ such that

$$f(c) = y.$$

Proof. Without loss of generality, assume that

$$f(a) < y < f(b).$$

The endpoint cases $y = f(a)$ or $y = f(b)$ are immediate in any case.

We now consider the set

$$S := \{x \in [a, b] : f(x) \leq y\}.$$

This set is nonempty, since $f(a) < y$ implies $a \in S$. Moreover, S is bounded above by b. Hence, by the completeness axiom, S has a supremum. We set

$$c := \sup S.$$

It remains to show that $f(c) = y$.

First we show that $f(c) \leq y$. Suppose, to the contrary, that $f(c) > y$. Then the continuity of f at c yields a $\delta > 0$ such that

$$f(x) > y$$

for all $x \in [a, b]$ with $|x - c| < \delta$. Since $c = \sup S$, however, there are points of S arbitrarily close to c from the left. In particular, there exists an $x \in S$ with

$$c - \delta < x \leq c.$$

For this x, on the one hand $x \in S$, so $f(x) \leq y$; on the other hand, $|x - c| < \delta$, so $f(x) > y$. This is a contradiction. Therefore $f(c) > y$ is impossible, and consequently

$$f(c) \leq y.$$

Now we show that $f(c) \geq y$. Suppose, to the contrary, that $f(c) < y$. Then the continuity of f at c yields a $\delta > 0$ such that

$$f(x) < y$$

for all $x \in [a, b]$ with $|x - c| < \delta$. Since $f(b) > y$, we have $b \notin S$. Hence $c < b$, and therefore there is still a part of the interval $[a, b]$ to the right of c. Thus there exists an $x \in [a, b]$ with

$$c < x < c + \delta.$$

For this x, the inequality $|x - c| < \delta$ gives $f(x) < y$, hence $x \in S$. But this contradicts the fact that c is an upper bound of S. Thus $f(c) < y$ is also impossible, and so

$$f(c) \geq y.$$

Together we obtain

$$f(c) \leq y \quad \text{and} \quad f(c) \geq y,$$

therefore

$$f(c) = y.$$

This proves the intermediate value theorem. □

Corollary 270 (Existence of a zero). *Let $f\colon [a,b] \to \mathbb{R}$ be a continuous function. If*

$$f(a) < 0 < f(b) \qquad \text{or} \qquad f(a) > 0 > f(b),$$

then there exists a point $y \in [a,b]$ such that

$$f(y) = 0.$$

Proof. In both cases, 0 lies between $f(a)$ and $f(b)$. Hence the intermediate value theorem yields a point $y \in [a,b]$ with $f(y) = 0$. □

Remark 271. *The intermediate value theorem is false in general if one considers the field $\mathbb{Q}$ instead of $\mathbb{R}$. A simple example is the function*

$$f\colon \mathbb{Q} \to \mathbb{Q}, \qquad f(x) = x^2.$$

It is continuous, but on the interval $[1,2] \cap \mathbb{Q}$ it does not attain the value 2. Indeed,

$$f(1) = 1 < 2 < 4 = f(2),$$

but there is no $c \in \mathbb{Q}$ such that

$$f(c) = 2.$$

Otherwise one would have

$$c^2 = 2$$

. *However, in an earlier chapter we showed that* $\sqrt{2}$ *is irrational. Therefore the equation*

$$x^2 = 2$$

has no rational solution.

Thus the intermediate value 2 is not attained by the continuous function $f(x) = x^2$ *on* $\mathbb{Q}$*. This shows that the intermediate value theorem depends essentially on the completeness of the real numbers and does not hold in the field of rational numbers.*

Example 272. *Consider the function*

$$f\colon [1,2] \to \mathbb{R}, \qquad f(x) = x^3 - x - 1.$$

Since f *is a polynomial function, it is continuous on all of* $\mathbb{R}$ *and in particular on* $[1,2]$*. Moreover,*

$$f(1) = 1 - 1 - 1 = -1 \qquad \text{and} \qquad f(2) = 8 - 2 - 1 = 5.$$

Thus the value 0 *lies between* $f(1)$ *and* $f(2)$*. By the intermediate value theorem, there therefore exists a*

$$c \in [1,2]$$

such that

$$f(c) = 0.$$

Hence the equation

$$x^3 - x - 1 = 0$$

has at least one real solution in the interval $[1,2]$*.*

The intermediate value theorem is therefore not merely a formal consequence of continuity, but one of the first fundamental existence theorems of analysis. We shall encounter it again and again later, for example when proving the existence of zeros and when studying continuous functions on intervals.

Corollary 273 (Continuous images of intervals are intervals)**.** *Let $I \subset \mathbb{R}$ be an interval, and let $f\colon I \to \mathbb{R}$ be continuous. Then*

$$f(I) \subset \mathbb{R}$$

is also an interval.

Proof. We show that $f(I)$ has the intermediate value property. Thus let $u, v \in f(I)$ with $u < v$, and let $w \in \mathbb{R}$ be a number such that

$$u < w < v.$$

We have to show that then also $w \in f(I)$ holds.

Since $u, v \in f(I)$, there exist points $x_1, x_2 \in I$ such that

$$f(x_1) = u \qquad \text{and} \qquad f(x_2) = v.$$

Without loss of generality, assume that $x_1 < x_2$. Since I is an interval, it follows that

$$[x_1, x_2] \subset I.$$

The restriction of f to the interval $[x_1, x_2]$ is therefore still continuous.

Now u and v lie on different sides of w. Hence the intermediate value theorem, applied to the continuous function $f|_{[x_1,x_2]}$, yields a point

$$c \in [x_1, x_2]$$

such that

$$f(c) = w.$$

Thus $w \in f(I)$ has been shown.

Therefore, between any two points of $f(I)$, every intermediate value again belongs to $f(I)$. Consequently, $f(I)$ is an interval. □

10.6 Bounded Functions and Extreme Values

Continuity has not only qualitative but also quantitative consequences. On a closed and bounded interval, a continuous function can neither take arbitrarily large values nor "miss" its greatest or smallest function value. This is precisely what the following section summarizes.

Definition 274. *A function $f\colon D \to \mathbb{R}$ is called* bounded *if its image $f(D)$ is bounded. That is, there exists an $A > 0$ such that*

$$|f(x)| \leq A \qquad \text{for all } x \in D.$$

Definition 275. *An interval is called* compact *if it is closed and bounded. In $\mathbb{R}$, these are precisely the intervals of the form*

$$[a, b] \qquad (a \leq b).$$

The next statement is one of the most important principles in analysis. It says that continuous functions on compact intervals behave particularly well: they are bounded and even attain their maximum and minimum.

Theorem 276 (Extreme value theorem). *Every continuous function*

$$f\colon [a,b] \to \mathbb{R}$$

is bounded. Moreover, there exist points $p, q \in [a,b]$ *such that*

$$\begin{aligned} f(p) &= \sup\{f(x) : x \in [a,b]\} \quad \textit{and} \\ f(q) &= \inf\{f(x) : x \in [a,b]\}. \end{aligned}$$

Thus the function f *attains its maximum and its minimum on* $[a,b]$.

Proof. We first show that f is bounded.

Suppose that f were not bounded above. Then for every $n \in \mathbb{N}$ there would exist an $x_n \in [a,b]$ such that

$$f(x_n) \geq n.$$

The sequence (x_n) lies in the bounded interval $[a,b]$. By the Bolzano-Weierstrass theorem, it therefore has a convergent subsequence (x_{n_k}) with limit

$$x_{n_k} \to c \in [a,b].$$

Since f is continuous, it follows that

$$f(x_{n_k}) \to f(c).$$

On the other hand, we also have

$$f(x_{n_k}) \geq n_k \to \infty,$$

so $(f(x_{n_k}))$ cannot have a finite limit. This is a contradiction. Hence f is bounded above.

In exactly the same way, one shows that f is also bounded below. Therefore f is bounded overall.

Now set

$$M := \sup\{f(x) : x \in [a, b]\}.$$

Because of boundedness, this supremum exists as a real number. We want to show that M is actually attained.

By the definition of the supremum, for every $n \in \mathbb{N}$ there exists an $x_n \in [a, b]$ such that

$$M - \frac{1}{n} < f(x_n) \leq M.$$

Again, the Bolzano-Weierstrass theorem yields a convergent subsequence (x_{n_k}) with

$$x_{n_k} \to c \in [a, b].$$

From the continuity of f it follows that

$$f(x_{n_k}) \to f(c).$$

On the other hand, since

$$M - \frac{1}{n_k} < f(x_{n_k}) \leq M$$

we also have

$$f(x_{n_k}) \to M.$$

Thus

$$f(c) = M.$$

Hence f attains its maximum.

For the minimum, one either considers analogously

$$m := \inf\{f(x) : x \in [a, b]\}$$

or applies what has already been shown to the continuous function $-f$. In either case, one obtains a $q \in [a, b]$ such that

$$f(q) = m.$$

Thus the minimum is also attained. □

The theorem shows that continuity on a compact interval is far more than merely local regularity: it prevents function values from escaping upward or downward and also prevents extrema from being missed.

Example 277. *The assumption of compactness is essential in the maximum and minimum theorem.*

Consider the continuous function

$$f\colon (0, 1) \to \mathbb{R}, \qquad f(x) = x.$$

Then

$$f((0, 1)) = (0, 1).$$

In particular, f is bounded. We have

$$\sup f((0, 1)) = 1 \qquad \textit{and} \qquad \inf f((0, 1)) = 0.$$

However, since neither 0 *nor* 1 *belongs to the interval* $(0, 1)$, *there is no point* $x \in (0, 1)$ *such that*

$$f(x) = 0 \qquad or \qquad f(x) = 1.$$

Thus the function has neither a minimum nor a maximum.

This example shows that merely omitting the endpoints can already cause a continuous function not to attain its extrema.

On other non-compact intervals, even boundedness may be lost. For instance, the continuous function

$$g \colon \mathbb{R} \to \mathbb{R}, \qquad g(x) = x,$$

is not bounded, since its values can become arbitrarily large and arbitrarily small.

The two examples show that both aspects of compactness are essential: boundedness of the interval prevents function values from escaping upward or downward, while closedness ensures that possible extrema at the endpoints can actually be attained.

Theorem 278 (ϵ-θ criterion for continuity). *Let* $D \subset \mathbb{R}$ *and let* $f \colon D \to \mathbb{R}$ *be a function. Then* f *is continuous at a point* $p \in D$ *if and only if for every* $\epsilon > 0$ *there exists a* $\theta > 0$ *such that*

$$|f(x) - f(p)| < \epsilon$$

for all $x \in D$ *satisfying*

$$|x - p| < \theta.$$

Proof. We prove the two directions separately.

"Only if": Suppose that f is continuous at p. We have to show that for every $\epsilon > 0$ there exists a $\theta > 0$ such that, whenever $x \in D$ and $|x - p| < \theta$, it follows that $|f(x) - f(p)| < \epsilon$.

Assume, to the contrary, that this is not the case. Then there would exist an $\epsilon_0 > 0$ such that for every $\theta > 0$ there is a point $x \in D$ with

$$|x - p| < \theta \qquad \text{and} \qquad |f(x) - f(p)| \geq \epsilon_0$$

exists. In particular, for each $n \in \mathbb{N}$ we can choose an $x_n \in D$ with

$$|x_n - p| < \frac{1}{n} \qquad \text{and} \qquad |f(x_n) - f(p)| \geq \epsilon_0.$$

Then $x_n \to p$. By the continuity of f at p, however, it follows that

$$f(x_n) \to f(p),$$

so for sufficiently large n one must have

$$|f(x_n) - f(p)| < \epsilon_0$$

holds. This contradicts the construction of the sequence (x_n). Hence the asserted ϵ-θ property must hold.

"If": Conversely, suppose that the ϵ-θ property holds. We show that f is then continuous at p.

Let (x_n) be a sequence in D with

$$x_n \to p.$$

We have to show that

$$f(x_n) \to f(p).$$

Let $\epsilon > 0$ be arbitrary. By assumption, there exists a $\theta > 0$ such that

$$|f(x) - f(p)| < \epsilon$$

for all $x \in D$ with $|x - p| < \theta$.

Since $x_n \to p$, there exists an $N \in \mathbb{N}$ such that

$$|x_n - p| < \theta$$

for all $n \geq N$. For these n, it follows that

$$|f(x_n) - f(p)| < \epsilon.$$

Since $\epsilon > 0$ was arbitrary, this proves that

$$f(x_n) \to f(p)$$

has been shown. Thus f is continuous at p.

This proves both directions. □

Corollary 279. *Let $f\colon D \to \mathbb{R}$ be continuous at $p \in D$, and suppose that*

$$f(p) \neq 0.$$

Then there exists a $\theta > 0$ such that

$$f(x) \neq 0$$

for all $x \in D$ with

$$|x - p| < \theta.$$

Proof. Set

$$\epsilon := \frac{|f(p)|}{2} > 0.$$

By the ϵ-θ criterion for continuity, there exists a $\theta > 0$ such that

$$|f(x) - f(p)| < \epsilon$$

for all $x \in D$ with $|x - p| < \theta$.

Suppose that, for such an x, one had $f(x) = 0$. Then

$$|f(x) - f(p)| = |f(p)| > \frac{|f(p)|}{2} = \epsilon,$$

which contradicts the choice of θ. Hence $f(x) \neq 0$ must hold. □

10.7 Uniform Continuity

Continuity of a function is initially a local property: for each point p, the appropriate θ may depend on that point. This is sufficient for many applications. In some situations, however, a stronger condition is desirable. One then asks not only whether the function behaves continuously near a single point, but whether the dependence between input and output values can be controlled throughout the entire domain by one and the same bound.

This leads precisely to the concept of uniform continuity. Whereas ordinary continuity allows a separate θ to be chosen for each point p, uniform continuity requires one single θ to work for all points of the domain. Thus uniform continuity is a global form of continuity.

Definition 280. *A function $f\colon D \to \mathbb{R}$ is called* uniformly continuous *on D if for every $\epsilon > 0$ there exists a $\theta > 0$ such that for all $x, y \in D$ with*

$$|x - y| < \theta$$

the inequality

$$|f(x) - f(y)| < \epsilon$$

holds.

The difference from ordinary continuity is therefore that θ here does not depend on a point $p \in D$, but only on ϵ. The control over the function values is thus uniform throughout the entire domain.

Example 281. *Every uniformly continuous function is, in particular, continuous at every point of its domain. Indeed, ordinary continuity follows immediately from the definition of uniform continuity by fixing one of the two points.*

The converse, however, is not true in general. A function may be continuous at every individual point without its continuity being controllable by a common θ on the whole domain. The following example illustrates this. The function

$$f\colon \mathbb{R} \to \mathbb{R}, \qquad f(x) = x^2,$$

is continuous at every point of $\mathbb{R}$*, but it is not uniformly continuous on* $\mathbb{R}$*.*

To see this, for $n \in \mathbb{N}$ *consider the two numbers*

$$x_n := n \qquad \text{and} \qquad y_n := n + \frac{1}{n}.$$

Then

$$|x_n - y_n| = \frac{1}{n} \to 0.$$

Thus the points become arbitrarily close to each other. For the function values, however, we obtain

$$\begin{aligned} |f(x_n) - f(y_n)| &= \left| n^2 - \left(n + \frac{1}{n}\right)^2 \right| \\ &= \left| n^2 - \left(n^2 + 2 + \frac{1}{n^2}\right) \right| \\ &= 2 + \frac{1}{n^2} \end{aligned}$$

Hence

$$|f(x_n) - f(y_n)| \to 2.$$

In particular, the function values do not become arbitrarily close to each other, even though the differences between the inputs become arbitrarily small. Therefore f *is not uniformly continuous.*

Theorem 282 (Heine-Cantor)**.** *Every function that is continuous on a compact interval* $[a, b]$*,*

$$f \colon [a, b] \to \mathbb{R}$$

is uniformly continuous on $[a, b]$*.*

Proof. We prove the statement by contradiction. Suppose that f were not uniformly continuous. Then there exists an $\epsilon_0 > 0$ such that for every $\theta > 0$ there are points $x, y \in [a, b]$ with

$$|x - y| < \theta \qquad \text{and} \qquad |f(x) - f(y)| \geq \epsilon_0$$

.

In particular, for every $n \in \mathbb{N}$ we can choose points $x_n, y_n \in [a, b]$ with

$$|x_n - y_n| < \frac{1}{n} \qquad \text{and} \qquad |f(x_n) - f(y_n)| \geq \epsilon_0.$$

Since both sequences (x_n) and (y_n) lie in the compact interval $[a, b]$, the Bolzano-Weierstrass theorem implies that the sequence (x_n) has a convergent subsequence (x_{n_k}). Write

$$x_{n_k} \to x \in [a, b].$$

From

$$|x_{n_k} - y_{n_k}| < \frac{1}{n_k} \to 0$$

it then also follows that

$$y_{n_k} \to x.$$

Indeed,

$$|y_{n_k} - x| \leq |y_{n_k} - x_{n_k}| + |x_{n_k} - x| \to 0.$$

Since f is continuous at every point of $[a, b]$, it follows from $x_{n_k} \to x$ and $y_{n_k} \to x$ that

$$f(x_{n_k}) \to f(x) \qquad \text{and} \qquad f(y_{n_k}) \to f(x).$$

Hence we also have

$$|f(x_{n_k}) - f(y_{n_k})| \to 0.$$

However, this contradicts the construction of the sequences, since for all k we had

$$|f(x_{n_k}) - f(y_{n_k})| \geq \epsilon_0.$$

This contradiction shows that our assumption was false. Therefore f is uniformly continuous on $[a, b]$. □

This theorem is one of the most important links between continuity and compactness: on a compact interval, continuity is not merely a local property, but can even be controlled uniformly.

10.7.1 Step Functions and Approximation of Continuous Functions

Before closing this chapter, we record an important structural property of continuous functions on compact intervals: such functions can be squeezed arbitrarily closely from below and from above by very simple functions. These simple functions are called *step functions.*

Definition 283. *A function $\varphi\colon [a, b] \to \mathbb{R}$ is called a* step function *if there is a partition*

$$a = x_0 < x_1 < \cdots < x_n = b$$

and numbers $c_1, \ldots, c_n \in \mathbb{R}$ such that

$$\varphi(x) = c_k \qquad \textit{for all } x \in [x_{k-1}, x_k)$$

for $k = 1, \ldots, n-1$, *and*

$$\varphi(x) = c_n \qquad \text{for all } x \in [x_{n-1}, x_n]$$

holds.

A step function is therefore constant on each subinterval of a finite partition. Its graph consists of finitely many horizontal "steps".

Theorem 284. *Let* $f\colon [a,b] \to \mathbb{R}$ *be a continuous function. Then the following holds: for every* $\epsilon > 0$ *there exist step functions*

$$\alpha, \beta\colon [a,b] \to \mathbb{R}$$

such that

$$\alpha(x) \leq f(x) \leq \beta(x) \qquad \text{for all } x \in [a,b]$$

and

$$|\alpha(x) - \beta(x)| \leq \epsilon \qquad \text{for all } x \in [a,b].$$

Proof. Since f is continuous on the compact interval $[a,b]$, the Heine–Cantor theorem implies that f is uniformly continuous. Hence, for the given $\epsilon > 0$, there exists a $\delta > 0$ such that

$$|f(x) - f(y)| < \epsilon$$

for all $x, y \in [a,b]$ with

$$|x - y| < \delta.$$

We now choose a partition

$$a = x_0 < x_1 < \cdots < x_n = b$$

of the interval $[a, b]$ so fine that, for every $k = 1, \ldots, n$,

$$x_k - x_{k-1} < \delta.$$

On each subinterval $[x_{k-1}, x_k]$, the function f is continuous; therefore, by the extreme value theorem, it attains both a minimum and a maximum there. We set

$$m_k := \min\{f(x) : x \in [x_{k-1}, x_k]\},$$
$$M_k := \max\{f(x) : x \in [x_{k-1}, x_k]\}.$$

Then, for all $x \in [x_{k-1}, x_k]$,

$$m_k \leq f(x) \leq M_k.$$

Moreover, the choice of the partition and uniform continuity imply that, for all $x, y \in [x_{k-1}, x_k]$,

$$|f(x) - f(y)| < \epsilon.$$

In particular, this also gives

$$M_k - m_k \leq \epsilon.$$

We now define step functions α and β by

$$\alpha(x) := m_k, \qquad \beta(x) := M_k$$

for $x \in [x_{k-1}, x_k)$, $k = 1, \ldots, n-1$, and on the last interval $[x_{n-1}, x_n]$ we likewise set

$$\alpha(x) := m_n, \qquad \beta(x) := M_n.$$

Then α and β are step functions.

For every $x \in [a, b]$, the point x lies in one of the subintervals $[x_{k-1}, x_k]$. Hence

$$\alpha(x) = m_k \leq f(x) \leq M_k = \beta(x).$$

Furthermore,

$$|\alpha(x) - \beta(x)| = M_k - m_k \leq \epsilon.$$

Thus both asserted properties have been proved. □

The theorem shows that continuous functions on compact intervals are not only regular, but can also be approximated in a very strong sense by simple piecewise constant functions. This is one important reason why continuous functions are so well manageable in analysis; later, this approximation will play a central role in particular in integration theory.

10.8 Exercises and Solutions

Exercise 285. *Show that the function*

$$f \colon \mathbb{R} \to \mathbb{R}, \qquad f(x) = \sqrt{x^2 + 1},$$

is continuous on all of $\mathbb{R}$.

Solution 286. *We decompose f into simpler continuous functions:*

$$x \mapsto x^2, \qquad x \mapsto x^2 + 1, \qquad x \mapsto \sqrt{x}.$$

The square function is continuous on $\mathbb{R}$, and hence the function $x \mapsto x^2+1$ is continuous as a sum of continuous functions. Since $x^2+1 \geq 1$ for all $x \in \mathbb{R}$, the square root function is defined on this range. Because the square root function is continuous on $[0, \infty)$, the composition

$$x \mapsto \sqrt{x^2+1}$$

is continuous on all of $\mathbb{R}$.

Exercise 287. *Investigate the equation*

$$x^3 - x - 1 = 0$$

for real solutions in the interval $[1, 2]$.

Solution 288. *Consider the function*

$$f(x) = x^3 - x - 1.$$

As a polynomial function, it is continuous on all of $\mathbb{R}$. Moreover,

$$f(1) = 1 - 1 - 1 = -1 \qquad \text{and} \qquad f(2) = 8 - 2 - 1 = 5.$$

Thus 0 lies between $f(1)$ and $f(2)$. By the intermediate value theorem, there therefore exists a point

$$c \in [1, 2]$$

such that

$$f(c) = 0.$$

Hence the equation has at least one real solution in $[1, 2]$.

Exercise 289. *Determine the maximum and the minimum of the function*

$$f\colon [0,1] \to \mathbb{R}, \qquad f(x) = x(1-x).$$

Solution 290. *We write*

$$f(x) = x - x^2.$$

Since f *is a polynomial, it is continuous on* $[0, 1]$. *By the maximum and minimum theorem,* f *attains both a maximum and a minimum there.*

First we consider the endpoint values:

$$f(0) = 0, \qquad f(1) = 0.$$

Moreover,

$$f(x) = x - x^2 = \frac{1}{4} - \left(x - \frac{1}{2}\right)^2.$$

Since always

$$\left(x - \frac{1}{2}\right)^2 \geq 0$$

holds, it follows that

$$f(x) \leq \frac{1}{4}$$

for all $x \in [0,1]$, with equality precisely for $x = \frac{1}{2}$. Thus the maximum is

$$\max_{x\in[0,1]} f(x) = \frac{1}{4}$$

at $x = \frac{1}{2}$.

Since $f(x) \geq 0$ on $[0,1]$, the minimum is

$$\min_{x\in[0,1]} f(x) = 0,$$

and it is attained at $x = 0$ and at $x = 1$.

Exercise 291. *Show that the function*

$$f\colon \mathbb{R} \to \mathbb{R}, \qquad f(x) = x^2,$$

is not uniformly continuous on $\mathbb{R}$.

Solution 292. *We show this using a sequence of pairs of points whose distance tends to 0, while the corresponding function values do not become arbitrarily close.*

For $n \in \mathbb{N}$, set

$$x_n := n \qquad \text{and} \qquad y_n := n + \frac{1}{n}.$$

Then

$$|x_n - y_n| = \frac{1}{n} \to 0.$$

On the other hand,

$$\begin{aligned}|f(x_n) - f(y_n)| &= \left|n^2 - \left(n + \frac{1}{n}\right)^2\right| \\ &= \left|n^2 - \left(n^2 + 2 + \frac{1}{n^2}\right)\right| \\ &= 2 + \frac{1}{n^2}.\end{aligned}$$

In particular,

$$|f(x_n) - f(y_n)| \to 2.$$

Thus, despite arbitrarily small differences in the input values, the function values do not become arbitrarily close. Hence there can be no common $\theta > 0$ that works for all points of $\mathbb{R}$. Therefore f is not uniformly continuous.

Exercise 293. *For every number $\epsilon > 0$, construct step functions $\alpha, \beta\colon [0,1] \to \mathbb{R}$ such that*

$$\alpha(x) \leq x \leq \beta(x) \qquad \text{and} \qquad \beta(x) - \alpha(x) \leq \epsilon$$

for all $x \in [0,1]$.

Solution 294. *First choose a natural number n such that*

$$\frac{1}{n} \leq \epsilon.$$

Divide the interval $[0,1]$ into the subintervals

$$\left[\frac{k}{n}, \frac{k+1}{n}\right) \qquad (k = 0, \ldots, n-1)$$

and the final interval

$$\left[\frac{n-1}{n}, 1\right].$$

Now define

$$\alpha(x) := \frac{k}{n}, \qquad \beta(x) := \frac{k+1}{n}$$

for $x \in \left[\frac{k}{n}, \frac{k+1}{n}\right)$, *and correspondingly on the final interval by*

$$\alpha(x) := \frac{n-1}{n}, \qquad \beta(x) := 1.$$

Then α *and* β *are step functions. Moreover, for every* $x \in [0,1]$ *we have*

$$\alpha(x) \leq x \leq \beta(x),$$

because x *lies in the corresponding subinterval. Finally, for all* $x \in [0,1]$,

$$\beta(x) - \alpha(x) = \frac{1}{n} \leq \epsilon.$$

Thus the required step functions have been constructed.

Chapter 11

Elementary Functions

In the preceding chapters, the focus was first on the foundations of analysis: completeness, sequences, series, point sets, and continuity. We now turn to a class of functions that reappears throughout almost all later parts of analysis: the *elementary functions.*

The term *elementary function* should not be understood here as a strictly formal mathematical category, but rather as a collective name for those functions that occur most frequently in practice and make later chapters especially transparent. These include, for example, power functions, root functions, exponential and logarithmic functions, trigonometric and inverse trigonometric functions, as well as composite and piecewise-defined functions. Especially in the chapters on differentiation, optimization problems, and integration, it will be essential to work with as broad a selection of such functions as possible.

It is particularly important that we regard functions not merely as formal rules, but as objects with geometric and analytic behavior: some functions increase, others decrease; some are invertible, others are not; some are defined on all of $\mathbb{R}$, others only on intervals; and still others arise from suitable combinations of simple building blocks. These properties will later allow us to interpret derivatives meaningfully, investigate extrema systematically, and compute integrals for concrete classes of functions.

11.1 Monotone Functions and Inverse Functions

We begin with one of the most fundamental properties of functions: monotonicity. Monotone functions form the starting point for the construction of inverse functions. In particular, for many elementary functions it will later be essential to determine on which intervals they are strictly monotone. Only then can we meaningfully invert them.

Definition 295 (Monotonicity). *Let $D \subset \mathbb{R}$ and let $f\colon D \to \mathbb{R}$ be a function.*

- *f is called* increasing *on D if, for all $x, y \in D$ with $x < y$, one has*

$$f(x) \leq f(y).$$

- *f is called* strictly increasing *on D if, for all $x, y \in D$ with $x < y$, one has*

$$f(x) < f(y).$$

- *f is called* decreasing *on D if, for all $x, y \in D$ with $x < y$, one has*

$$f(x) \geq f(y).$$

- *f is called* strictly decreasing *on D if, for all $x, y \in D$ with $x < y$, one has*

$$f(x) > f(y).$$

Increasing and decreasing functions are often grouped together under the term monotone; *strictly increasing and strictly decreasing functions are called* strictly monotone.

Definition 296 (Inverse function). *Let $f\colon D \to \mathbb{R}$ be a function. If f is injective, then the corresponding inverse mapping*

$$f^{-1}\colon f(D) \to D$$

is called the inverse function *of f. It is uniquely determined by the condition*

$$f^{-1}(b) = a \quad \Longleftrightarrow \quad f(a) = b$$

for all $a \in D$ and $b \in f(D)$.

Theorem 297 (Continuity of the inverse function). *Let $D \subset \mathbb{R}$ be an interval and let*

$$f\colon D \to \mathbb{R}$$

be a continuous, strictly monotone function.

Then $E := f(D)$ is again an interval, and f maps D bijectively onto E. The inverse function

$$f^{-1}\colon E \to D$$

is continuous.

Proof. Since f is continuous and D is an interval, the image

$$E = f(D)$$

is also an interval by the result of the previous chapter.

Because f is strictly monotone, it is injective: if, for example, $x < y$, then in the strictly increasing case one has $f(x) < f(y)$, while in the strictly decreasing case one has $f(x) > f(y)$. Thus two distinct points of D cannot have the same function value. Hence $f\colon D \to E$ is bijective, and the inverse function $f^{-1}\colon E \to D$ is well defined.

It remains to prove the continuity of f^{-1}. We give the proof for the case where f is strictly increasing; the strictly decreasing case is completely analogous.

Let $b \in E$ be given, and set

$$a := f^{-1}(b).$$

We show that f^{-1} is continuous at b. To this end, let $\varepsilon > 0$ be given.

Case 1: a is an interior point of D. Then we may choose ε so small that

$$[a-\varepsilon, a+\varepsilon] \subset D.$$

Set

$$b_1 := f(a-\varepsilon), \qquad b_2 := f(a+\varepsilon).$$

By the strict monotonicity of f, we have

$$b_1 < b < b_2.$$

Since f is continuous and strictly monotone on the interval $[a-\varepsilon, a+\varepsilon]$, it maps this interval bijectively onto the interval $[b_1, b_2]$.

Now let

$$\theta := \min\{\, b - b_1,\; b_2 - b \,\}.$$

Then $\theta > 0$, and for every $y \in E$ with

$$|y - b| < \theta$$

it follows that

$$y \in (b_1, b_2).$$

Therefore the preimage $f^{-1}(y)$ also lies in the interval $(a-\varepsilon, a+\varepsilon)$, and hence

$$|f^{-1}(y) - a| < \varepsilon.$$

Since for every $\varepsilon > 0$ a corresponding $\theta > 0$ has been found, the inverse function f^{-1} is continuous at b by the ε-θ criterion.

Case 2: b is an endpoint of E. Then $a = f^{-1}(b)$ is correspondingly an endpoint of D. If, for instance, b is the left endpoint of E, then instead of a symmetric interval we consider only a one-sided neighbourhood of a, that is, an interval of the form $[a, a + \varepsilon]$; if b is the right endpoint of E, we argue analogously with $[a - \varepsilon, a]$.

Since f is again continuous and strictly monotone on the subinterval under consideration, we obtain the same conclusion as in the interior case: small changes of b produce only small changes of $f^{-1}(b)$. Thus f^{-1} is continuous at endpoints as well.

This proves the continuity of f^{-1} at every point of E. □

Remark 298. *This theorem is of great importance for the further development of the theory. Later, we shall introduce many elementary functions by considering a known continuous and strictly monotone function on a suitable interval and then defining its inverse function. In this way, for example, root functions, logarithms, and the inverse functions of trigonometric functions arise.*

Theorem 299 (Monotonicity of the inverse function). *Let $D \subset \mathbb{R}$ be an interval and let*

$$f\colon D \to \mathbb{R}$$

be a strictly monotone function.

Then the inverse function

$$f^{-1}\colon f(D) \to D$$

is also strictly monotone. If f is strictly increasing, then f^{-1} is strictly increasing as well; if f is strictly decreasing, then f^{-1} is strictly decreasing as well.

Proof. We first consider the case in which f is strictly increasing.

Let $b_1, b_2 \in f(D)$ with $b_1 < b_2$, and set

$$a_1 := f^{-1}(b_1), \qquad a_2 := f^{-1}(b_2).$$

Then

$$f(a_1) = b_1 < b_2 = f(a_2).$$

If we had $a_1 \geq a_2$, then the strict monotonicity of f would imply

$$f(a_1) \geq f(a_2),$$

contradicting $f(a_1) < f(a_2)$. Hence $a_1 < a_2$, and therefore f^{-1} is strictly increasing.

In the case where f is strictly decreasing, the argument is analogous: from $b_1 < b_2$ it follows immediately, by the strict decreasingness of f, that $a_1 > a_2$. Thus f^{-1} is strictly decreasing. □

11.2 Power and Root Functions

Root functions are among the first nontrivial inverse functions that play a central role in analysis. They connect power functions with a new and important class of functions and will reappear frequently later in the study of limits, derivatives, optimization problems, and integrals.

In what follows, let $k \in \mathbb{N}$ with $k \geq 1$ be fixed. We consider the power function

$$p_k : \mathbb{R}_+ \to \mathbb{R}_+, \qquad p_k(x) = x^k,$$

where $\mathbb{R}_+ := (0, \infty)$ denotes the set of positive real numbers.

Lemma 300 (Strict monotonicity of the power function). *For every $k \in \mathbb{N}$, the function*

$$p_k : \mathbb{R}_+ \to \mathbb{R}_+, \qquad x \mapsto x^k$$

is strictly increasing.

Proof. Let $x, y \in \mathbb{R}_+$ with $x < y$. Then

$$y^k - x^k = (y - x)\big(y^{k-1} + y^{k-2}x + \cdots + yx^{k-2} + x^{k-1}\big).$$

The first factor is positive, and every summand in the parentheses is positive as well. Hence

$$y^k - x^k > 0,$$

so $x^k < y^k$. Therefore, p_k is strictly increasing. □

Lemma 301 (Bijectivity of the power function). *For every $k \in \mathbb{N}$, the function*

$$p_k : \mathbb{R}_+ \to \mathbb{R}_+$$

is bijective.

Proof. Injectivity already follows from strict monotonicity.

To prove surjectivity, let $y \in \mathbb{R}_+$ be arbitrary. We show that there exists an $x \in \mathbb{R}_+$ such that $x^k = y$.

First choose $a > 0$ large enough that $a^k > y$. This is possible because $a^k \to \infty$ as $a \to \infty$. On the other hand, for sufficiently small $b > 0$ we have $b^k < y$. Since p_k is continuous on the interval $[b, a]$, the intermediate value theorem implies that p_k takes every value between b^k and a^k, in particular the value y.

Thus there exists an $x \in [b, a] \subset \mathbb{R}_+$ with $x^k = y$. Hence p_k is surjective and therefore bijective. □

Definition 302 (k-th root function). *The inverse function of*

$$p_k \colon \mathbb{R}_+ \to \mathbb{R}_+, \qquad x \to x^k$$

is called the k-th root function *and is denoted by*

$$\sqrt[k]{\cdot} \colon \mathbb{R}_+ \to \mathbb{R}_+, \qquad x \mapsto \sqrt[k]{x}$$

.

Theorem 303 (Continuity of the k-th root function). *For every $k \in \mathbb{N}$, the k-th root function*

$$\sqrt[k]{\cdot} \colon \mathbb{R}_+ \to \mathbb{R}_+$$

is continuous.

Proof. By the preceding lemmas, the function $p_k(x) = x^k$ is continuous, strictly increasing, and bijective from $\mathbb{R}_+$ onto $\mathbb{R}_+$.

Thus it satisfies the assumptions of the theorem on the continuity of the inverse function for strictly monotone functions. Its inverse function is therefore continuous. This inverse function is precisely the k-th root function,

$$p_k^{-1}(x) = \sqrt[k]{x}.$$

Hence $\sqrt[k]{\cdot}$ is continuous as well. □

Theorem 304 (Monotonicity of the k-th root function). *For every $k \in \mathbb{N}$, the k-th root function*

$$\sqrt[k]{\cdot} : \mathbb{R}_+ \to \mathbb{R}_+$$

is strictly increasing.

Proof. The function $p_k(x) = x^k$ is strictly increasing on $\mathbb{R}_+$. By the theorem on the monotonicity of inverse functions, its inverse function is therefore also strictly increasing. Since the inverse function of p_k is precisely the k-th root function, it follows that

$$x < y \quad \Longrightarrow \quad \sqrt[k]{x} < \sqrt[k]{y}$$

for all $x, y \in \mathbb{R}_+$. □

Remark 305. *For $k = 2$, we obtain the square root*

$$\sqrt{x} = \sqrt[2]{x},$$

for $k = 3$ the cube root, and in general the k-th root. Root functions will be important later, especially in the study of power laws, limits, and derivatives.

11.3 Exponential and Logarithmic Functions

After having studied inverse functions in general, we now turn to two of the most important elementary functions: the exponential function and the natural logarithm. Both functions will be used repeatedly in the following chapters, especially in connection with limits, derivatives, growth behavior, and integrals.

In what follows, we denote by

$$\exp\colon \mathbb{R} \to \mathbb{R}_{>0}$$

the exponential function. We assume that exp is continuous, strictly increasing, and bijective. Its inverse function

$$\log\colon \mathbb{R}_{>0} \to \mathbb{R}$$

is called the *natural logarithm.*

Theorem 306 (Continuity of the natural logarithm). *The function*

$$\log\colon \mathbb{R}_{>0} \to \mathbb{R}$$

is continuous.

Proof. By assumption, the exponential function $\exp\colon \mathbb{R} \to \mathbb{R}_{>0}$ is continuous, strictly increasing, and bijective. Hence it is a continuous strictly monotone function on an interval, namely on all of $\mathbb{R}$.

By the theorem on the continuity of inverse functions, its inverse function is also continuous. This inverse function is precisely the natural logarithm

$$\log\colon \mathbb{R}_{>0} \to \mathbb{R}.$$

Therefore log is continuous. □

Theorem 307 (Monotonicity of the natural logarithm). *The function*

$$\log\colon \mathbb{R}_{>0} \to \mathbb{R}$$

is strictly increasing.

Proof. The exponential function $\exp\colon \mathbb{R} \to \mathbb{R}_{>0}$ is strictly increasing. By the theorem on the monotonicity of inverse functions, its inverse function is therefore also strictly increasing.

Since the inverse function of exp is the natural logarithm, it follows that

$$\log\colon \mathbb{R}_{>0} \to \mathbb{R}$$

is strictly increasing. □

Theorem 308 (Logarithmic law for products). *For all $x, y \in \mathbb{R}_{>0}$, one has*

$$\log(xy) = \log(x) + \log(y).$$

Proof. Let $x, y \in \mathbb{R}_{>0}$. Since the exponential function is bijective, there exist unique numbers $u, v \in \mathbb{R}$ such that

$$x = \exp(u), \qquad y = \exp(v).$$

Thus

$$xy = \exp(u)\exp(v).$$

For the exponential function, the calculation rule

$$\exp(u+v) = \exp(u)\exp(v).$$

holds. Hence

$$xy = \exp(u+v).$$

Applying the natural logarithm to both sides, we obtain

$$\log(xy) = \log(\exp(u+v)) = u+v.$$

Since $u = \log(x)$ and $v = \log(y)$, this is precisely

$$\log(xy) = \log(x) + \log(y).$$

□

11.3.1 Exponential function with base a

We have already encountered the natural exponential function

$$\exp\colon \mathbb{R} \to \mathbb{R}_{>0}$$

and the natural logarithm

$$\log\colon \mathbb{R}_{>0} \to \mathbb{R}$$

With their help, we can now define exponential functions for arbitrary positive bases.

Definition 309 (Exponential function with base a). *Let $a > 0$. We define the exponential function with base a by*

$$\exp_a : \mathbb{R} \to \mathbb{R}_{>0}, \qquad \exp_a(x) := \exp(x \log(a)).$$

We also write

$$\exp_a(x) = a^x.$$

Remark 310. *For $a = 1$, one has $1^x = 1$ for all $x \in \mathbb{R}$. If $a \neq 1$, the function $x \mapsto a^x$ is strictly monotone: it is strictly increasing for $a > 1$ and strictly decreasing for $0 < a < 1$.*

Theorem 311 (Basic algebraic law of the exponential function). *For all $a > 0$ and all $x, y \in \mathbb{R}$, we have*

$$\exp_a(x + y) = \exp_a(x) \exp_a(y).$$

Proof. By definition, we have

$$\exp_a(x+y) = \exp((x+y)\log(a)) = \exp(x \log(a) + y \log(a)).$$

For the natural exponential function, we already know that

$$\exp(u + v) = \exp(u) \exp(v) \qquad \text{for all } u, v \in \mathbb{R}.$$

Therefore,

$$\exp_a(x+y) = \exp(x \log(a)) \exp(y \log(a)) = \exp_a(x) \exp_a(y).$$

□

Theorem 312 (Integer exponents). *For every $a > 0$ and every $n \in \mathbb{Z}$, one has*

$$\exp_a(n) = a^n.$$

Proof. First let $n \in \mathbb{N}_0$. We prove the statement by induction.

For $n = 0$, we have

$$\exp_a(0) = \exp(0) = 1 = a^0.$$

Assume that $\exp_a(n) = a^n$ already holds. Then the product rule gives

$$\exp_a(n+1) = \exp_a(n)\exp_a(1) = a^n \cdot \exp_a(1).$$

Now

$$\exp_a(1) = \exp(\log(a)) = a,$$

hence

$$\exp_a(n+1) = a^n \cdot a = a^{n+1}.$$

Thus the assertion holds for all $n \in \mathbb{N}_0$.

For negative integers $n = -m$ with $m \in \mathbb{N}$, we obtain

$$\begin{aligned} 1 = \exp_a(0) &= \exp_a(m + (-m)) \\ &= \exp_a(m)\exp_a(-m) = a^m \exp_a(-m). \end{aligned}$$

Therefore

$$\exp_a(-m) = \frac{1}{a^m} = a^{-m}.$$

Consequently, for all $n \in \mathbb{Z}$,

$$\exp_a(n) = a^n.$$

□

Theorem 313 (Rational exponents). *For every $a > 0$, every $p \in \mathbb{Z}$, and every $q \in \mathbb{N}$ with $q \geq 2$, one has*

$$\exp_a\left(\frac{p}{q}\right) = \sqrt[q]{a^p}.$$

Proof. Set

$$x := \exp_a\left(\frac{p}{q}\right).$$

Then the product rule, applied q times, yields

$$x^q = \left(\exp_a\left(\frac{p}{q}\right)\right)^q = \exp_a\left(q \cdot \frac{p}{q}\right) = \exp_a(p).$$

By the preceding theorem,

$$\exp_a(p) = a^p.$$

Hence

$$x^q = a^p.$$

Since $x > 0$, the number x is precisely the positive q-th root of a^p, and therefore

$$x = \sqrt[q]{a^p}.$$

Thus

$$\exp_a\left(\frac{p}{q}\right) = \sqrt[q]{a^p}.$$

□

Remark 314. *The formula*

$$\exp_a(x) = \exp(x \log(a))$$

is particularly useful because it allows the properties of the exponential function with base a to be derived directly from the already known properties of exp *and* log. *In this way, continuity, monotonicity, and differentiation rules can later be obtained very elegantly.*

Theorem 315 (Further rules for powers). *Let $a, b > 0$ and $x, y \in \mathbb{R}$. Then:*

1. $$a^{x+y} = a^x a^y.$$

2. $$a^{x-y} = \frac{a^x}{a^y}.$$

3. $$(a^x)^y = a^{xy}.$$

4. $$(ab)^x = a^x b^x.$$

5. $$\left(\frac{1}{a}\right)^x = a^{-x}.$$

Proof. We use the definition

$$a^x = \exp_a(x) = \exp(x \log(a)) \qquad (a > 0,\ x \in \mathbb{R}).$$

For 1. We have

$$a^{x+y} = \exp((x+y)\log(a)) = \exp(x\log(a) + y\log(a)).$$

Using the product rule for the natural exponential function, it follows that

$$\exp(x\log(a)+y\log(a)) = \exp(x\log(a))\exp(y\log(a)) = a^x a^y.$$

For 2. By part 1 we obtain

$$a^x = a^{(x-y)+y} = a^{x-y}a^y.$$

Since $a^y > 0$, we may divide by a^y and obtain

$$a^{x-y} = \frac{a^x}{a^y}.$$

For 3. We write

$$(a^x)^y = \exp(y\log(a^x)).$$

Since $a^x = \exp(x\log(a))$ and the natural logarithm is the inverse function of exp, it follows that

$$\log(a^x) = \log(\exp(x\log(a))) = x\log(a).$$

Thus

$$(a^x)^y = \exp(y\,x\log(a)) = \exp((xy)\log(a)) = a^{xy}.$$

For 4. First,

$$a^x b^x = \exp(x\log(a))\exp(x\log(b)) = \exp(x(\log(a)+\log(b))).$$

By the logarithm law for products,

$$\log(a) + \log(b) = \log(ab).$$

Therefore,

$$a^x b^x = \exp(x \log(ab)) = (ab)^x.$$

For 5. First,

$$\left(\frac{1}{a}\right)^x = \exp\left(x \log\left(\frac{1}{a}\right)\right).$$

Furthermore,

$$\log\left(\frac{1}{a}\right) = \log(1) - \log(a) = -\log(a),$$

because $\log(1) = 0$. Hence

$$\left(\frac{1}{a}\right)^x = \exp(-x \log(a)) = a^{-x}.$$

□

Remark 316. *With these rules, we can perform calculations with powers in exactly the same way as in the case of integer exponents. In particular, the familiar laws of exponents are now extended to all real exponents. This will be of central importance later for limits, differentiation, and integrals.*

Theorem 317. *For every $a > 0$, one has*

$$\lim_{n\to\infty} \sqrt[n]{a} = 1.$$

Proof. We write the n-th root by means of the exponential function with base e:

$$\sqrt[n]{a} = a^{\frac{1}{n}} = \exp\left(\frac{1}{n}\log(a)\right).$$

Set $c := \log(a) \in \mathbb{R}$. Then the expression becomes

$$\sqrt[n]{a} = \exp\left(\frac{c}{n}\right).$$

Since

$$\lim_{n\to\infty} \frac{c}{n} = 0,$$

and the exponential function $\exp\colon \mathbb{R} \to \mathbb{R}_{>0}$ is continuous, the limit theorem for continuous functions yields

$$\lim_{n\to\infty} \exp\left(\frac{c}{n}\right) = \exp\left(\lim_{n\to\infty} \frac{c}{n}\right) = \exp(0) = 1.$$

Thus we have shown that

$$\lim_{n\to\infty} \sqrt[n]{a} = 1.$$

□

Remark 318. *This result shows in particular that powers of the form $a^{1/n}$ converge to 1 independently of the value of $a > 0$. This is a central limiting behavior of the exponential function and will later play an important role in growth estimates and series.*

Theorem 319 (Characterization of continuous multiplicative functions). *Let*

$$f: \mathbb{R} \to \mathbb{R}$$

be a continuous function satisfying

$$f(x+y) = f(x)f(y) \qquad \forall\, x, y \in \mathbb{R}.$$

Then either

$$f(x) = 0 \qquad \forall\, x \in \mathbb{R},$$

or

$$a := f(1) > 0$$

and

$$f(x) = a^x \qquad \forall\, x \in \mathbb{R}.$$

Proof. From the functional equation with $x = y = \frac{1}{2}$, we obtain

$$f(1) = f\left(\frac{1}{2} + \frac{1}{2}\right) = f\left(\frac{1}{2}\right)^2 \geq 0.$$

Thus only the two cases $f(1) = 0$ and $f(1) > 0$ can occur.

Case 1: $f(1) = 0$. Then, for every $x \in \mathbb{R}$,

$$f(x) = f(1 + (x-1)) = f(1)f(x-1) = 0.$$

Hence f is identically zero.

Case 2: $f(1) > 0$. Set

$$a := f(1).$$

First, from

$$f(1) = f(1+0) = f(1)f(0)$$

and since $f(1) \neq 0$, it follows immediately that

$$f(0) = 1.$$

We now show that, on the rational numbers, f agrees with a power function.

For $n \in \mathbb{N}_0$, complete induction using the functional equation gives

$$f(n) = a^n.$$

Indeed,

$$f(0) = 1 = a^0,$$

and if $f(n) = a^n$ holds, then

$$f(n+1) = f(n+1 \cdot 1) = f(n)f(1) = a^n a = a^{n+1}.$$

For negative integers $-n$ with $n \in \mathbb{N}$, we have

$$1 = f(0) = f(n+(-n)) = f(n)f(-n) = a^n f(-n),$$

hence

$$f(-n) = a^{-n}.$$

Therefore,

$$f(n) = a^n \qquad \forall\, n \in \mathbb{Z}.$$

Next we consider rational numbers. Let $p \in \mathbb{Z}$ and $q \in \mathbb{N}$ with $q \geq 2$. Then

$$f(p) = f\left(q \cdot \frac{p}{q}\right) = f\left(\frac{p}{q}\right)^q.$$

Using the already established result $f(p) = a^p$, it follows that

$$f\left(\frac{p}{q}\right)^q = a^p.$$

Since $f\left(\frac{p}{q}\right)$ cannot be negative, by continuity and the functional equation, and is uniquely determined in the positive case, we obtain

$$f\left(\frac{p}{q}\right) = \sqrt[q]{a^p}.$$

Thus, for all rational numbers $r \in \mathbb{Q}$,

$$f(r) = a^r.$$

Now let $x \in \mathbb{R}$ be arbitrary. By the density of the rational numbers, there exists a sequence $(r_n)_{n\in\mathbb{N}}$ of rational numbers such that

$$r_n \to x.$$

Since f is continuous, it follows that

$$f(x) = \lim_{n\to\infty} f(r_n).$$

Because $f(r_n) = a^{r_n}$ for all n and the exponential function with base a is continuous, we obtain

$$f(x) = \lim_{n\to\infty} a^{r_n} = a^x.$$

This proves that

$$f(x) = a^x \qquad \forall\, x \in \mathbb{R}.$$

□

Remark 320. *This characterization explains why the definition*

$$a^x := \exp(x \log(a))$$

for $a > 0$ is the natural one: a continuous function satisfying the fundamental rule

$$f(x + y) = f(x)f(y)$$

must either be identically zero or be precisely such a power function. In particular, this automatically gives

$$a^{x+y} = a^x a^y \qquad \forall\, x, y \in \mathbb{R}$$

and the power notation describes not merely a formal expression, but exactly the corresponding functional behavior.

11.4 Limits of Power and Logarithmic Functions

The introduction of exponential and logarithmic functions becomes truly meaningful once we know some of their fundamental limits. These limits will be needed repeatedly later on in comparisons of growth, in sequences and series, and in differential and integral calculus. What is especially important is that most of these statements follow directly from continuity, monotonicity, and the algebraic rules of the exponential and logarithmic functions.

Theorem 321 (Limit of the power function near 0). *For every $a > 0$, one has*

$$\lim_{x \to 0} a^x = 1.$$

Proof. By the definition of the power function with base a, we have

$$a^x = \exp(x \log(a)).$$

Since $x \mapsto x \log(a)$ is continuous and converges to 0 as $x \to 0$, the continuity of the exponential function implies

$$\lim_{x \to 0} a^x = \lim_{x \to 0} \exp(x \log(a)) = \exp(0) = 1.$$

□

Theorem 322 (Behavior of the power function for large arguments). *Let $a > 0$.*

- *If $a > 1$, then*

$$\lim_{x \to \infty} a^x = \infty.$$

- *If $0 < a < 1$, then*

$$\lim_{x \to \infty} a^x = 0.$$

Proof. Again we write

$$a^x = \exp(x \log(a)).$$

Case 1: $a > 1$. Then $\log(a) > 0$. Hence

$$x \log(a) \to \infty \qquad (x \to \infty).$$

Since the exponential function is strictly increasing and diverges to ∞, it follows that

$$a^x = \exp(x \log(a)) \to \infty.$$

Case 2: $0 < a < 1$. Then $\log(a) < 0$. Hence

$$x \log(a) \to -\infty \qquad (x \to \infty).$$

Since $\exp(t) \to 0$ as $t \to -\infty$, it follows that

$$a^x = \exp(x \log(a)) \to 0.$$

□

Theorem 323 (Behavior of the natural logarithm). *We have*

$$\lim_{x\to\infty} \log(x) = \infty \qquad \textit{and} \qquad \lim_{x\to 0+} \log(x) = -\infty.$$

Proof. Since log is the inverse function of the strictly increasing exponential function, log is also strictly increasing.

For $x \to \infty$. Let $M \in \mathbb{R}$ be arbitrary. Then

$$\log(\exp(M)) = M.$$

By the strict monotonicity of log, it follows immediately from $x > \exp(M)$ that

$$\log(x) > M.$$

Thus $\log(x)$ becomes arbitrarily large as $x \to \infty$, that is,

$$\log(x) \to \infty.$$

For $x \to 0^+$. Let $M \in \mathbb{R}$ be arbitrary. Then $\exp(M) > 0$. For every x with

$$0 < x < \exp(M)$$

the strict monotonicity of log gives

$$\log(x) < M.$$

Therefore,

$$\log(x) \to -\infty \qquad (x \to 0^+).$$

□

Theorem 324 (The exponential function dominates linear growth). *Let* $r > 1$. *Then*

$$\lim_{x\to\infty} \frac{x}{r^x} = 0.$$

Proof. Since $r > 1$, we can choose a natural number m such that

$$r > 1 + \frac{1}{m}.$$

By Bernoulli's inequality, it follows that

$$\left(1 + \frac{1}{m}\right)^m \geq 2.$$

Now let $x > 0$ and set

$$n := \lfloor x/m \rfloor.$$

Then

$$mn \leq x < m(n+1).$$

Since $r > 1$, it follows from $x \geq mn$ that

$$r^x \geq r^{mn}.$$

Moreover,

$$r^{mn} > \left(1 + \frac{1}{m}\right)^{mn} = \left(\left(1 + \frac{1}{m}\right)^m\right)^n \geq 2^n.$$

Hence we obtain

$$0 \leq \frac{x}{r^x} \leq \frac{m(n+1)}{2^n}.$$

Since $\frac{n+1}{2^n} \to 0$ as $n \to \infty$, it follows that

$$\frac{x}{r^x} \to 0 \qquad (x \to \infty).$$

□

Remark 325. *Many useful limits follow immediately from these theorems. For example, for $a > 1$ one has*

$$\lim_{x\to\infty} \frac{x}{a^x} = 0,$$

and from the substitution $x = \exp(t)$ it follows for every $\alpha > 0$ that

$$\lim_{x\to\infty} \frac{\log(x)}{x^\alpha} = 0.$$

Such estimates will play an important role later in comparisons of growth and in questions of convergence.

11.5 Landau Notation

In the study of limits, it is often more practical not only to state the exact limit of a function, but also to describe its relative growth behavior. For this purpose one uses *Landau notation.* It is especially useful when comparing functions with one another, for example powers, logarithms, and exponential functions.

In what follows, $x \to \infty$ is always meant unless stated otherwise.

Definition 326 (Landau symbols). *Let $f, g\colon (a, \infty) \to \mathbb{R}$ be functions, and suppose that $g(x) \neq 0$ for all sufficiently large x.*

- *One writes*

  $$f(x) = O(g(x)) \qquad (x \to \infty),$$

  if there exist constants $C > 0$ and $x_0 > a$ such that

  $$|f(x)| \leq C|g(x)| \qquad \forall\, x \geq x_0.$$

- *One writes*

  $$f(x) = o(g(x)) \qquad (x \to \infty),$$

  if

  $$\lim_{x\to\infty} \frac{f(x)}{g(x)} = 0.$$

- *One writes*

$$f(x) \sim g(x) \qquad (x \to \infty),$$

if

$$\lim_{x\to\infty} \frac{f(x)}{g(x)} = 1.$$

Remark 327. *Thus the notation $f(x) = O(g(x))$ describes an estimate of order of magnitude, while $f(x) = o(g(x))$ states that f vanishes in comparison with g. Finally, the relation $f \sim g$ says that the two functions are asymptotically of the same size.*

Theorem 328. *One has*

$$\log(x) = o(x^\alpha) \qquad (x \to \infty)$$

for every $\alpha > 0$.

Proof. Set $x = \mathrm{e}^t$ with $t \to \infty$. Then

$$\frac{\log(x)}{x^\alpha} = \frac{t}{\exp(\alpha t)}.$$

From the preceding section we know that exponential functions dominate every linear growth, hence

$$\lim_{t\to\infty} \frac{t}{\exp(\alpha t)} = 0.$$

It follows that

$$\lim_{x\to\infty} \frac{\log(x)}{x^\alpha} = 0,$$

and therefore

$$\log(x) = o(x^\alpha).$$

□

Theorem 329. *For every $a > 1$ and every $\alpha > 0$, one has*

$$x^\alpha = o(a^x) \qquad (x \to \infty).$$

Proof. Again write $a^x = \exp(x \log a)$. Then

$$\frac{x^\alpha}{a^x} = \frac{x^\alpha}{\exp(x \log a)}.$$

The exponential function in the denominator grows faster than any power in the numerator. Hence

$$\lim_{x \to \infty} \frac{x^\alpha}{a^x} = 0.$$

Thus

$$x^\alpha = o(a^x).$$

□

Theorem 330. *For every $a > 0$, one has*

$$a^x = 1 + o(1) \qquad (x \to 0).$$

Proof. From the already proved continuity of the exponential function, it follows that

$$\lim_{x \to 0} a^x = 1.$$

Hence

$$a^x - 1 \to 0,$$

and therefore

$$a^x = 1 + o(1).$$

□

Remark 331. *Typical notations are, for example,*

$$\log(x) = o(x^\alpha), \qquad x^\alpha = o(a^x), \qquad a^x = 1 + o(1).$$

Such formulas are very compact and will be useful later when we want to express limits and error terms in a shorter notation.

11.6 Exercises and Solutions

Exercise 332. *Show that for every $k \in \mathbb{N}$ and all $x, y \in \mathbb{R}_+$ with $x < y$, one has*

$$x^k < y^k.$$

Solution 333. *We use the well-known factorization*

$$y^k - x^k = (y - x)(y^{k-1} + y^{k-2}x + \cdots + yx^{k-2} + x^{k-1}).$$

Since $x < y$ and both numbers are positive, the factor $y - x$ is positive, and each summand in the parentheses is positive as well. Thus

$$y^k - x^k > 0,$$

hence

$$x^k < y^k.$$

Therefore, the function $x \mapsto x^k$ *is strictly increasing on* $\mathbb{R}_+$.

Exercise 334. *Justify that the function*

$$p_k \colon \mathbb{R}_+ \to \mathbb{R}_+, \qquad p_k(x) = x^k$$

is bijective.

Solution 335. *Injectivity already follows from the strict monotonicity established in the preceding exercise.*

To prove surjectivity, let $y \in \mathbb{R}_+$ *be arbitrary. We consider the function* $p_k(x) = x^k$. *For sufficiently small* $b > 0$ *we have* $b^k < y$, *and for sufficiently large* $a > 0$ *we have* $a^k > y$. *Since* p_k *is continuous, the intermediate value theorem implies that on the interval* $[b, a]$ *it attains every value between* b^k *and* a^k. *In particular, there exists an* $x \in [b, a] \subset \mathbb{R}_+$ *such that*

$$x^k = y.$$

Thus p_k *is surjective and hence bijective.*

Exercise 336. *Show that the k-th root function*

$$\sqrt[k]{\cdot} \colon \mathbb{R}_+ \to \mathbb{R}_+$$

is strictly increasing.

Solution 337. *The k-th root function is the inverse function of the strictly increasing function*

$$p_k(x) = x^k \qquad (x \in \mathbb{R}_+).$$

By the theorem on the monotonicity of inverse functions, it follows that

$$\sqrt[k]{\cdot}$$

is also strictly increasing. In other words, from $0 < x < y$ *it follows that*

$$\sqrt[k]{x} < \sqrt[k]{y}.$$

Exercise 338. *Prove the calculation rule*

$$a^{x+y} = a^x a^y.$$

for all $a > 0$ *and all* $x, y \in \mathbb{R}$.

Solution 339. *Using the definition*

$$a^x = \exp(x \log(a))$$

we obtain

$$a^{x+y} = \exp((x + y) \log(a)) = \exp(x \log(a) + y \log(a)).$$

Using the corresponding calculation rule for the exponential function, it follows that

$$\exp(x \log(a) + y \log(a)) = \exp(x \log(a)) \exp(y \log(a)) = a^x a^y.$$

This proves the formula.

Exercise 340. *Show that for every* $a > 0$,

$$\lim_{n\to\infty} \sqrt[n]{a} = 1.$$

Solution 341. *We have*

$$\sqrt[n]{a} = a^{1/n} = \exp\left(\frac{1}{n}\log(a)\right).$$

Since

$$\frac{1}{n}\log(a) \to 0 \qquad (n \to \infty)$$

and the exponential function is continuous, it follows that

$$\lim_{n\to\infty} \sqrt[n]{a} = \exp(0) = 1.$$

Exercise 342. *Show that for every* $\alpha > 0$,

$$\log(x) = o(x^\alpha) \qquad (x \to \infty).$$

Solution 343. *Set* $x = \mathrm{e}^t$, *so that* $t = \log(x)$ *and* $t \to \infty$ *as* $x \to \infty$. *Then*

$$\frac{\log(x)}{x^\alpha} = \frac{t}{\exp(\alpha t)}.$$

From the previous section we know that the exponential function dominates every linear growth, hence

$$\lim_{t\to\infty} \frac{t}{\exp(\alpha t)} = 0.$$

Therefore,

$$\lim_{x\to\infty} \frac{\log(x)}{x^\alpha} = 0,$$

and thus

$$\log(x) = o(x^\alpha).$$

Exercise 344. *Show that for every $a > 1$ and every $\alpha > 0$,*

$$x^\alpha = o(a^x) \qquad (x \to \infty).$$

Solution 345. *We write*

$$a^x = \exp(x \log(a)).$$

Since $a > 1$, we have $\log(a) > 0$. Thus the exponent $x \log(a)$ grows linearly to ∞, whereas the numerator is only a power. Therefore,

$$\lim_{x\to\infty} \frac{x^\alpha}{a^x} = \lim_{x\to\infty} \frac{x^\alpha}{\exp(x \log(a))} = 0.$$

Hence,

$$x^\alpha = o(a^x).$$

11.7 Trigonometric Functions

After introducing the exponential and logarithmic functions, we now turn to another central class of elementary functions: the trigonometric functions. They are among the most important functions in analysis and arise naturally in geometry, physics, and later also in the theory of differential equations.

Unlike the functions considered so far, trigonometric functions are not primarily motivated by algebraic rules of calculation, but by a geometric construction. The starting point is the unit circle. This geometric origin explains many of their later properties, in particular their periodicity, their symmetries, and their characteristic values at special points.

At first, the focus is on the sine and cosine functions. From them, further trigonometric functions such as tangent and cotangent are then derived. We deliberately introduce sine and cosine without using the complex exponential function, and initially regard them as geometrically defined quantities before studying them as functions on $\mathbb{R}$.

A central goal is to understand the trigonometric functions as well-defined, continuous, and periodic functions, and to derive their fundamental properties from their definition.

11.7.1 The Unit Circle and Radian Measure

We begin with a basic geometric idea.

The *unit circle* is the circle in the plane with center at the origin and radius 1. Its equation is

$$x^2 + y^2 = 1.$$

The point $(1, 0)$ is chosen as the starting point. If this point is rotated on the unit circle by an angle counterclockwise, a new point on the circle is obtained. The coordinates of this point depend only on the angle of rotation.

In order for this procedure to correspond naturally to real numbers, we use not degrees, but *radian measure*.

Definition 346 (Radian measure). *An angle is measured in* radians *by assigning to it the length of the corresponding arc on the unit circle.*

Since the circumference of the unit circle is equal to 2π, one full revolution corresponds to the angle 2π. Half a revolution therefore corresponds to the angle π.

Definition 347 (Sine and cosine via the unit circle). *Let $x \in \mathbb{R}$. We consider the point on the unit circle obtained from the point $(1,0)$ by a rotation through the angle x. If this point has coordinates*

$$P(x) = (u, v),$$

then we define

$$\cos x := u \qquad \textit{and} \qquad \sin x := v.$$

Thus sine and cosine are defined as functions

$$\sin, \cos : \mathbb{R} \to \mathbb{R}$$

.

Geometrically, the cosine is read as the horizontal coordinate and the sine as the vertical coordinate of the corresponding point on the unit circle. From this definition, the first fundamental properties follow immediately.

Theorem 348 (Fundamental properties of sine and cosine). *For all $x \in \mathbb{R}$, one has*

$$\sin^2 x + \cos^2 x = 1.$$

Moreover, sin *and* cos *are periodic with period* 2π*, that is,*

$$\sin(x + 2\pi) = \sin x, \qquad \cos(x + 2\pi) = \cos x.$$

Proof. The point $P(x) = (\cos x, \sin x)$ lies on the unit circle. Hence it satisfies the equation of the unit circle:

$$(\cos x)^2 + (\sin x)^2 = 1.$$

Since a rotation through 2π returns to the starting point, the coordinates remain unchanged after one full revolution. Therefore sine and cosine are 2π-periodic. □

Remark 349. *The definition via the unit circle is particularly useful because it does not require complex numbers and at the same time makes the later algebraic laws intuitively clear. Properties such as symmetry, sign behavior, and periodicity can be understood directly from the geometry and will be proved rigorously in what follows.*

The figure shows the unit circle centered at the origin. If the radius starting at the point $(1, 0)$ is rotated counterclockwise by the angle x, one obtains the point $P(x)$ with

$$P(x) = (\cos x, \sin x).$$

Thus the coordinates of $P(x)$ are given directly by cosine and sine: $\cos x$ describes the horizontal position, while $\sin x$ describes the vertical position of the point. The dashed lines illustrate these projections onto the x- and y-axes.

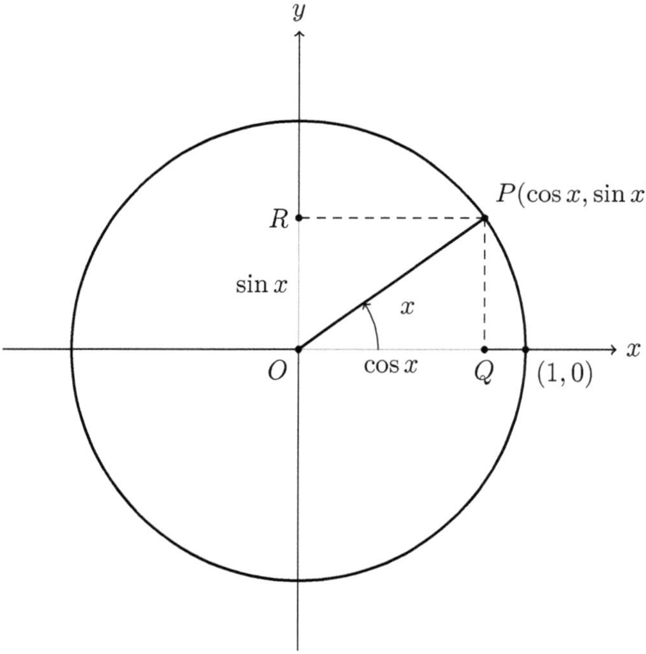

Figure 11.1: Unit circle: definition of sine and cosine

Theorem 350. *For all $x \in \mathbb{R}$, one has:*

$$\cos(-x) = \cos(x), \qquad \sin(-x) = -\sin(x).$$

Proof. The statement follows directly from the definition of sine and cosine on the unit circle. The point obtained from the angle x is mapped to the point corresponding to the angle $-x$ by reflection across the x-axis. In this process, the x-coordinate, that is, the cosine value, remains unchanged, whereas the sign of the y-coordinate, that is, the sine value, is reversed. □

Theorem 351 (Addition theorems for sine and cosine). *For all $x, y \in \mathbb{R}$, we have*

$$\cos(x+y) = \cos x \cos y - \sin x \sin y$$

and

$$\sin(x+y) = \sin x \cos y + \cos x \sin y.$$

Proof. We consider the unit circle and the point

$$P(x) = (\cos x, \sin x).$$

If this point is rotated counterclockwise about the origin by the angle y, the resulting point is $P(x+y)$.

Now rotation by the angle y is a length- and angle-preserving map. Hence the unit vectors

$$e_1 = (1,0), \qquad e_2 = (0,1)$$

have, after the rotation, the images

$$e_1' = (\cos y, \sin y) \quad \text{and} \quad e_2' = (-\sin y, \cos y).$$

Indeed, e_1' lies on the unit circle at angle y, and e_2' is obtained from it by a rotation through 90°.

The point $P(x)$ can be written in the standard basis as

$$P(x) = \cos x\, e_1 + \sin x\, e_2.$$

After rotation by y, this becomes

$$P(x+y) = \cos x\, e_1' + \sin x\, e_2'.$$

Substitution gives

$$P(x+y) = \cos x\,(\cos y, \sin y) + \sin x\,(-\sin y, \cos y).$$

Thus the coordinates of $P(x+y)$ are given by

$$P(x+y) = \bigl(\cos x \cos y - \sin x \sin y,\ \sin x \cos y + \cos x \sin y\bigr).$$

But since, at the same time,

$$P(x+y) = (\cos(x+y), \sin(x+y))$$

holds, comparison of coordinates yields precisely

$$\cos(x+y) = \cos x \cos y - \sin x \sin y$$

and

$$\sin(x+y) = \sin x \cos y + \cos x \sin y.$$

□

Theorem 352. *The functions*

$$\sin\colon \mathbb{R} \to \mathbb{R} \qquad \textit{and} \qquad \cos\colon \mathbb{R} \to \mathbb{R}$$

are continuous on all of $\mathbb{R}$.

Proof. Let $x_0 \in \mathbb{R}$ be arbitrary and let $h \in \mathbb{R}$. Using the addition formulas, we obtain

$$\sin(x_0 + h) - \sin(x_0) = \sin(x_0)(\cos h - 1) + \cos(x_0)\sin h,$$

as well as

$$\cos(x_0 + h) - \cos(x_0) = \cos(x_0)(\cos h - 1) - \sin(x_0)\sin h.$$

As $h \to 0$, we have $\sin h \to 0$ and $\cos h \to 1$. Hence both right-hand sides tend to 0, and therefore

$$\lim_{h\to 0} \sin(x_0{+}h) = \sin(x_0) \qquad \text{and} \qquad \lim_{h\to 0} \cos(x_0{+}h) = \cos(x$$

Thus sin and cos are continuous at x_0. Since x_0 was arbitrary, both functions are continuous on all of $\mathbb{R}$. □

Theorem 353 (Product-to-sum formulas). *For all $x, y \in \mathbb{R}$, we have*

$$\sin x - \sin y = 2\cos\left(\frac{x+y}{2}\right)\sin\left(\frac{x-y}{2}\right)$$

and

$$\cos x - \cos y = -2\sin\left(\frac{x+y}{2}\right)\sin\left(\frac{x-y}{2}\right).$$

Proof. Set

$$u = \frac{x+y}{2} \qquad \text{and} \qquad v = \frac{x-y}{2}.$$

Then

$$x = u + v \qquad \text{and} \qquad y = u - v.$$

Using the addition formulas for sine and cosine, we first obtain

$$\sin x - \sin y = \sin(u+v) - \sin(u-v).$$

Now

$$\sin(u+v) = \sin u \cos v + \cos u \sin v$$

and

$$\sin(u-v) = \sin u \cos v - \cos u \sin v.$$

Therefore, by subtraction,

$$\sin(u+v) - \sin(u-v) = 2\cos u \sin v.$$

Using the definitions of u and v, it follows that

$$\sin x - \sin y = 2\cos\left(\frac{x+y}{2}\right)\sin\left(\frac{x-y}{2}\right).$$

For cosine, we likewise use the addition formulas

$$\cos(u+v) = \cos u \cos v - \sin u \sin v$$

and

$$\cos(u-v) = \cos u \cos v + \sin u \sin v.$$

Thus

$$\cos(u+v) - \cos(u-v) = -2\sin u \sin v.$$

Again using $u = \frac{x+y}{2}$ and $v = \frac{x-y}{2}$, we obtain

$$\cos x - \cos y = -2\sin\left(\frac{x+y}{2}\right)\sin\left(\frac{x-y}{2}\right).$$

This proves both formulas. □

Theorem 354 (Absolute convergence of the sine and cosine series). *For every $x \in \mathbb{R}$, the series*

$$\sin x = \sum_{k=0}^{\infty} (-1)^k \frac{x^{2k+1}}{(2k+1)!}$$

and $$\cos x = \sum_{k=0}^{\infty} (-1)^k \frac{x^{2k}}{(2k)!}$$

converge absolutely.

Proof. We prove the absolute convergence of both series by the ratio test.

First we consider the sine series. For the absolute values of its terms, we set

$$a_k := \left| \frac{x^{2k+1}}{(2k+1)!} \right| = \frac{|x|^{2k+1}}{(2k+1)!}.$$

Then, for $k \geq 0$, we have

$$\frac{a_{k+1}}{a_k} = \frac{|x|^{2k+3}}{(2k+3)!} \cdot \frac{(2k+1)!}{|x|^{2k+1}} = \frac{|x|^2}{(2k+2)(2k+3)}.$$

For fixed x, this gives

$$\lim_{k \to \infty} \frac{|x|^2}{(2k+2)(2k+3)} = 0.$$

Since this limit is less than 1, the series

$$\sum_{k=0}^{\infty} a_k = \sum_{k=0}^{\infty} \frac{|x|^{2k+1}}{(2k+1)!}$$

converges by the ratio test. Hence the sine series is absolutely convergent.

Now we consider the cosine series. With

$$b_k := \left| \frac{x^{2k}}{(2k)!} \right| = \frac{|x|^{2k}}{(2k)!}$$

we obtain analogously

$$\frac{b_{k+1}}{b_k} = \frac{|x|^{2k+2}}{(2k+2)!} \cdot \frac{(2k)!}{|x|^{2k}} = \frac{|x|^2}{(2k+1)(2k+2)}.$$

Again,

$$\lim_{k\to\infty} \frac{|x|^2}{(2k+1)(2k+2)} = 0.$$

Therefore,

$$\sum_{k=0}^{\infty} b_k = \sum_{k=0}^{\infty} \frac{|x|^{2k}}{(2k)!}$$

also converges, so the cosine series is absolutely convergent as well.

Thus both series are absolutely convergent for every $x \in \mathbb{R}$. □

Corollary 355. *We have*

$$\lim_{x\to 0,\, x\neq 0} \frac{\sin x}{x} = 1 \qquad \textit{and} \qquad \lim_{x\to 0,\, x\neq 0} \frac{\cos x - 1}{x} = 0.$$

Proof. We consider the functions defined by the power series

$$\sin x = \sum_{k=0}^{\infty} (-1)^k \frac{x^{2k+1}}{(2k+1)!}, \qquad \cos x = \sum_{k=0}^{\infty} (-1)^k \frac{x^{2k}}{(2k)!}.$$

For the sine series, we write

$$\sin x = x + R(x),$$

where

$$R(x) = \sum_{k=1}^{\infty} (-1)^k \frac{x^{2k+1}}{(2k+1)!}$$

is the remainder of the series. Then, for $x \neq 0$,

$$\left|\frac{\sin x}{x} - 1\right| = \left|\frac{R(x)}{x}\right| \leq \sum_{k=1}^{\infty} \frac{|x|^{2k}}{(2k+1)!}.$$

For $|x| \leq 1$, this gives

$$\sum_{k=1}^{\infty} \frac{|x|^{2k}}{(2k+1)!} \leq |x|^2 \sum_{k=1}^{\infty} \frac{1}{(2k+1)!},$$

and the right-hand side tends to 0 as $x \to 0$. Hence

$$\frac{\sin x}{x} \to 1.$$

For the cosine series, we write

$$\cos x = 1 + S(x),$$

where

$$S(x) = \sum_{k=1}^{\infty} (-1)^k \frac{x^{2k}}{(2k)!}.$$

Then, for $x \neq 0$,

$$\left|\frac{\cos x - 1}{x}\right| = \left|\frac{S(x)}{x}\right| \leq \sum_{k=1}^{\infty} \frac{|x|^{2k-1}}{(2k)!},$$

and, for $|x| \leq 1$,

$$\sum_{k=1}^{\infty} \frac{|x|^{2k-1}}{(2k)!} \leq |x| \sum_{k=1}^{\infty} \frac{1}{(2k)!}.$$

The right-hand side tends to 0 as $x \to 0$. Therefore

$$\frac{\cos x - 1}{x} \to 0.$$

□

11.8 The Number π

Lemma 356 (Sign of sin). *For all $x \in (0, 2]$, we have*

$$\sin x > 0.$$

Proof. We write the sine series as an alternating series:

$$\sin x = x - \frac{x^3}{3!} + \frac{x^5}{5!} - \frac{x^7}{7!} + \cdots .$$

For $0 < x \leq 2$, the sequence of absolute values

$$a_k := \frac{x^{2k+1}}{(2k+1)!}$$

is strictly decreasing, since

$$\frac{a_{k+1}}{a_k} = \frac{x^2}{(2k+2)(2k+3)} \leq \frac{4}{6} = \frac{2}{3} < 1.$$

Thus the series is alternating with decreasing terms. By the Leibniz criterion, its value lies between two consecutive partial sums; in particular,

$$\sin x \geq x - \frac{x^3}{3!}.$$

Moreover, for $0 < x \leq 2$,

$$x - \frac{x^3}{6} = x\left(1 - \frac{x^2}{6}\right) > 0$$

holds. Hence, altogether,

$$\sin x > 0.$$

□

Lemma 357 (Monotonicity of cos). *The function* cos *is strictly decreasing on the interval* $[0, 2]$.

Proof. Let $x, y \in [0, 2]$ with $x < y$. Set

$$u := \frac{x+y}{2}, \qquad v := \frac{y-x}{2}.$$

Then

$$u \in (0, 2] \qquad \text{and} \qquad v \in (0, 1].$$

By the preceding lemma, therefore,

$$\sin u > 0 \qquad \text{and} \qquad \sin v > 0.$$

Using the addition theorem for cosine, we obtain

$$\cos y - \cos x = \cos(u + v) - \cos(u - v) = -2 \sin u \sin v.$$

Since both sine values are positive, we have

$$\cos y - \cos x < 0.$$

Thus

$$\cos y < \cos x,$$

and hence cos is strictly decreasing on $[0, 2]$. □

Theorem 358 (Existence and uniqueness of a zero of cosine). *The function* cos *has exactly one zero in* $(0, 2)$.

Proof. Since cos is continuous and

$$\cos 0 = 1 > 0 \qquad \text{and} \qquad \cos 2 < 0$$

hold, the intermediate value theorem yields the existence of a zero $c \in (0, 2)$ such that

$$\cos c = 0.$$

Because cos is strictly monotone on $[0, 2]$, there can be at most one zero there. Hence this zero is unique. □

Definition 359. *We set*

$$\pi := 2c,$$

where c *is the unique zero of* cos *in* $(0, 2)$. *Thus*

$$\cos\left(\frac{\pi}{2}\right) = 0.$$

Corollary 360. *We have*

$$\sin\left(\frac{\pi}{2}\right) = 1.$$

Proof. The point on the unit circle corresponding to the angle $\pi/2$ has the coordinates

$$\left(\cos\left(\frac{\pi}{2}\right), \sin\left(\frac{\pi}{2}\right)\right) = (0, \sin(\pi/2)).$$

Since this point lies on the unit circle, we have

$$\cos^2\left(\frac{\pi}{2}\right) + \sin^2\left(\frac{\pi}{2}\right) = 1.$$

Because $\cos(\pi/2) = 0$, it follows that

$$\sin^2\left(\frac{\pi}{2}\right) = 1.$$

Moreover, $\sin x > 0$ for $x \in (0, 2]$. Therefore

$$\sin\left(\frac{\pi}{2}\right) = 1.$$

$\square$

Theorem 361 (Symmetries and periodicity of the trigonometric functions). *For all $x \in \mathbb{R}$, the following identities hold:*

$$\cos(\pi + x) = -\cos x, \qquad \sin(\pi + x) = -\sin x,$$

$$\cos(x + 2\pi) = \cos x, \qquad \sin(x + 2\pi) = \sin x,$$

and also

$$\cos x = \sin\left(\frac{\pi}{2} - x\right), \qquad \sin x = \cos\left(\frac{\pi}{2} - x\right).$$

Proof. First, recall the values already obtained:

$$\cos\left(\frac{\pi}{2}\right) = 0, \qquad \sin\left(\frac{\pi}{2}\right) = 1.$$

The addition theorem then gives:

$$\cos(\pi) = \cos\left(\frac{\pi}{2} + \frac{\pi}{2}\right) = \cos^2\left(\frac{\pi}{2}\right) - \sin^2\left(\frac{\pi}{2}\right) = 0 - 1 = -1,$$

and

$$\sin(\pi) = \sin\left(\frac{\pi}{2} + \frac{\pi}{2}\right) = 2\sin\left(\frac{\pi}{2}\right)\cos\left(\frac{\pi}{2}\right) = 2 \cdot 1 \cdot 0 = 0.$$

Thus, for every $x \in \mathbb{R}$, the addition theorems yield:

$$\cos(\pi+x) = \cos\pi\ \cos x - \sin\pi\ \sin x = (-1)\cos x - 0 \cdot \sin x = -\mathrm{cc}$$

and

$$\sin(\pi+x) = \sin\pi\ \cos x + \cos\pi\ \sin x = 0 \cdot \cos x + (-1)\sin x = -\mathrm{si}$$

Applying these identities once more, we obtain

$$\begin{aligned} \cos(x + 2\pi) &= \cos(\pi + (\pi + x)) = -\cos(\pi + x) \\ &= -(-\cos x) = \cos x. \end{aligned}$$

and similarly

$$\begin{aligned} \sin(x + 2\pi) &= \sin(\pi + (\pi + x)) = -\sin(\pi + x) \\ &= -(-\sin x) = \sin x. \end{aligned}$$

This proves the 2π-periodicity.

For the cofunction identities, we substitute $u = \frac{\pi}{2}$ and $v = -x$ into the addition theorem. This gives

$$\begin{aligned}\cos\left(\frac{\pi}{2} - x\right) &= \cos\left(\frac{\pi}{2}\right)\cos x + \sin\left(\frac{\pi}{2}\right)\sin x\\ &= 0 \cdot \cos x + 1 \cdot \sin x\\ &= \sin x\end{aligned}$$

Hence

$$\sin x = \cos\left(\frac{\pi}{2} - x\right).$$

Analogously, the sine addition theorem gives

$$\begin{aligned}\sin\left(\frac{\pi}{2} - x\right) &= \sin\left(\frac{\pi}{2}\right)\cos x - \cos\left(\frac{\pi}{2}\right)\sin x\\ &= 1 \cdot \cos x - 0 \cdot \sin x\\ &= \cos x\end{aligned}$$

Thus

$$\cos x = \sin\left(\frac{\pi}{2} - x\right)$$

is also proved. □

Theorem 362 (Zeros of sine and cosine). *We have*

$$\{x \in \mathbb{R} : \sin x = 0\} = \{k\pi : k \in \mathbb{Z}\}$$

and

$$\{x \in \mathbb{R} : \cos x = 0\} = \left\{\frac{\pi}{2} + k\pi : k \in \mathbb{Z}\right\}.$$

Proof. We first consider the sine function. Because of its 2π-periodicity, it suffices to determine the zeros in an interval of length 2π. From the preceding theorems we know that

$$\sin(\pi + x) = -\sin x.$$

Together with the already established positivity of $\sin x$ on the first interval between two zeros, it follows that the sine function cannot change sign between two consecutive multiples of π, and that in each such interval there is exactly one zero at the boundary. Hence the zeros of sin are precisely the numbers

$$k\pi, \qquad k \in \mathbb{Z}.$$

For the cosine function we use the cofunction identity

$$\cos x = \sin\left(\frac{\pi}{2} - x\right).$$

This immediately gives

$$\cos x = 0 \iff \sin\left(\frac{\pi}{2} - x\right) = 0.$$

By the already proved zero theorem for the sine function, this is the case precisely when

$$\frac{\pi}{2} - x = k\pi \qquad \text{for some } k \in \mathbb{Z}.$$

Thus

$$x = \frac{\pi}{2} - k\pi.$$

Since k is an arbitrary integer, this is the same set as

$$\left\{\frac{\pi}{2} + k\pi : k \in \mathbb{Z}\right\}.$$

This describes both sets of zeros. $\square$

Definition 363 (Tangent). *For all $x \in \mathbb{R}$ with $\cos x \neq 0$, we define*

$$\tan x := \frac{\sin x}{\cos x}.$$

In particular, tan *is well-defined on the interval $\left(-\frac{\pi}{2}, \frac{\pi}{2}\right)$.*

Theorem 364 (Monotonicity and bijections of the trigonometric functions). *The following statements hold:*

1. *The function*

$$\sin\colon \left[-\frac{\pi}{2}, \frac{\pi}{2}\right] \to [-1, 1]$$

is strictly increasing and bijective.

2. *The function*

$$\cos\colon [0, \pi] \to [-1, 1]$$

is strictly decreasing and bijective.

3. *The function*

$$\tan\colon \left(-\frac{\pi}{2}, \frac{\pi}{2}\right) \to \mathbb{R}$$

is strictly increasing and bijective.

Proof. We prove the three statements one after another.

1. The sine function. Let $x, y \in \left[-\frac{\pi}{2}, \frac{\pi}{2}\right]$ with $x < y$. Set

$$u := \frac{x+y}{2}, \qquad v := \frac{y-x}{2}.$$

Then

$$u \in \left(-\frac{\pi}{2}, \frac{\pi}{2}\right), \qquad v \in \left(0, \frac{\pi}{2}\right].$$

By the addition theorem, we obtain

$$\sin y - \sin x = 2\cos(u)\sin(v).$$

It remains to show that both factors are positive.

Since cos has no zeros on $\left(-\frac{\pi}{2}, \frac{\pi}{2}\right)$ and $\cos 0 = 1 > 0$, the continuity of cos implies that

$$\cos(u) > 0.$$

Moreover,

$$\sin(v) = \cos\left(\frac{\pi}{2} - v\right),$$

and since $\frac{\pi}{2} - v \in \left[0, \frac{\pi}{2}\right)$, we also have

$$\sin(v) > 0.$$

Thus $\sin y - \sin x > 0$. Hence sin is strictly increasing.

Furthermore,

$$\sin\left(-\frac{\pi}{2}\right) = -1 \qquad \text{and} \qquad \sin\left(\frac{\pi}{2}\right) = 1,$$

and therefore, by continuity and strict monotonicity,

$$\sin\left(\left[-\frac{\pi}{2},\frac{\pi}{2}\right]\right)=[-1,1].$$

Thus sin is bijective from $\left[-\frac{\pi}{2},\frac{\pi}{2}\right]$ onto $[-1,1]$.

2. The cosine function. Let $x,y\in[0,\pi]$ with $x<y$. Again set

$$u:=\frac{x+y}{2},\qquad v:=\frac{y-x}{2}.$$

Then

$$u\in(0,\pi),\qquad v\in\left(0,\frac{\pi}{2}\right].$$

By the addition theorem,

$$\cos y-\cos x=-2\sin(u)\sin(v).$$

Now

$$\sin(u)=\cos\left(\frac{\pi}{2}-u\right),$$

and since $\frac{\pi}{2}-u\in\left(-\frac{\pi}{2},\frac{\pi}{2}\right)$, the same argument as above gives

$$\sin(u)>0.$$

Similarly, $\sin(v)>0$. Hence

$$\cos y-\cos x<0.$$

Thus cos is strictly decreasing on $[0,\pi]$.

Moreover,

$$\cos(0)=1\qquad\text{and}\qquad\cos(\pi)=-1.$$

By continuity and strict monotonicity, it follows that

$$\cos([0, \pi]) = [-1, 1].$$

Thus $\cos\colon [0, \pi] \to [-1, 1]$ is bijective.

3. The tangent function. On the interval $(-\frac{\pi}{2}, \frac{\pi}{2})$ we have $\cos x > 0$, so

$$\tan x = \frac{\sin x}{\cos x}$$

is well-defined and continuous there.

Now let $x, y \in (-\frac{\pi}{2}, \frac{\pi}{2})$ with $x < y$. Then

$$\tan y - \tan x = \frac{\sin y}{\cos y} - \frac{\sin x}{\cos x} = \frac{\sin y \cos x - \sin x \cos y}{\cos x \, \cos y}.$$

By the addition theorem,

$$\sin y \cos x - \sin x \cos y = \sin(y - x).$$

Hence

$$\tan y - \tan x = \frac{\sin(y - x)}{\cos x \, \cos y}.$$

Since

$$0 < y - x < \pi,$$

we have $\sin(y - x) > 0$. In addition, $\cos x$ and $\cos y$ are positive. Therefore,

$$\tan y - \tan x > 0.$$

Thus tan is strictly increasing.

For bijectivity it remains to show that tan assumes all real values. Since

$$\lim_{x\to\frac{\pi}{2}^-} \cos x = 0^+ \qquad \text{and} \qquad \lim_{x\to\frac{\pi}{2}^-} \sin x = 1,$$

it follows that

$$\lim_{x\to\frac{\pi}{2}^-} \tan x = +\infty.$$

By the oddness of tan, we similarly obtain

$$\lim_{x\to-\frac{\pi}{2}^+} \tan x = -\infty.$$

Since tan is continuous and strictly increasing, it assumes every real value exactly once. Hence

$$\tan\colon \left(-\frac{\pi}{2}, \frac{\pi}{2}\right) \to \mathbb{R}$$

is bijective. □

Definition 365 (Inverse trigonometric functions)**.** *The inverse functions of the restricted trigonometric functions are defined as follows:*

$$\arcsin\colon [-1,1] \to \left[-\frac{\pi}{2}, \frac{\pi}{2}\right],$$

$$\arccos\colon [-1,1] \to [0,\pi],$$

$$\arctan\colon \mathbb{R} \to \left(-\frac{\pi}{2}, \frac{\pi}{2}\right).$$

Thus,

$$\arcsin(\sin x) = x \quad \textit{for } x \in \left[-\frac{\pi}{2}, \frac{\pi}{2}\right],$$
$$\arccos(\cos x) = x \quad \textit{for } x \in [0, \pi],$$
$$\arctan(\tan x) = x \quad \textit{for } x \in \left(-\frac{\pi}{2}, \frac{\pi}{2}\right).$$

11.9 Exercises and Solutions

Exercise 366. *Determine the following values:*

$$\sin(0), \quad \cos(0), \quad \sin\left(\frac{\pi}{2}\right), \quad \cos\left(\frac{\pi}{2}\right),$$
$$\sin(\pi), \quad \cos(\pi), \quad \sin(2\pi), \quad \cos(2\pi).$$

Solution 367. *From the properties already proved, it follows that*

$$\sin(0) = 0, \qquad \cos(0) = 1.$$

Moreover,

$$\sin\left(\frac{\pi}{2}\right) = 1, \qquad \cos\left(\frac{\pi}{2}\right) = 0.$$

Using the formulas

$$\sin(\pi + x) = -\sin x, \qquad \cos(\pi + x) = -\cos x$$

we obtain

$$\sin(\pi) = 0, \qquad \cos(\pi) = -1.$$

Finally, the 2π-periodicity and the values at 0 yield

$$\sin(2\pi) = 0, \qquad \cos(2\pi) = 1.$$

Exercise 368. *Use the addition theorems to show that*

$$\sin(2x) = 2\sin x\cos x \qquad \text{and} \qquad \cos(2x) = \cos^2 x - \sin^2 x.$$

Solution 369. *By the addition theorems, we have*

$$\sin(2x) = \sin(x+x) = \sin x\cos x + \cos x\sin x = 2\sin x\cos x.$$

Similarly,

$$\cos(2x) = \cos(x+x) = \cos x\cos x - \sin x\sin x = \cos^2 x - \sin^2 x$$

Exercise 370. *Determine all solutions of*

$$\sin x = 0 \qquad \text{and} \qquad \cos x = 0$$

in $\mathbb{R}$.

Solution 371. *The zeros of the sine function are precisely the integer multiples of π:*

$$\sin x = 0 \iff x = k\pi \quad (k \in \mathbb{Z}).$$

The zeros of the cosine function are precisely the numbers

$$x = \frac{\pi}{2} + k\pi \quad (k \in \mathbb{Z}).$$

Thus,

$$\{x \in \mathbb{R} : \sin x = 0\} = \{k\pi : k \in \mathbb{Z}\}$$

and

$$\{x \in \mathbb{R} : \cos x = 0\} = \left\{\frac{\pi}{2} + k\pi : k \in \mathbb{Z}\right\}.$$

Exercise 372. *Solve the equations on the indicated intervals:*

$$\sin x = \frac{1}{2} \quad \textit{for } x \in \left[-\frac{\pi}{2}, \frac{\pi}{2}\right],$$

$$\cos x = \frac{1}{2} \quad \textit{for } x \in [0, \pi],$$

$$\tan x = 1 \quad \textit{for } x \in \left(-\frac{\pi}{2}, \frac{\pi}{2}\right).$$

Solution 373. *Since* arcsin *is the inverse function of* sin *on* $\left[-\frac{\pi}{2}, \frac{\pi}{2}\right]$, *it follows that*

$$\sin x = \frac{1}{2} \iff x = \arcsin\left(\frac{1}{2}\right) = \frac{\pi}{6}.$$

Since arccos *is the inverse function of* cos *on* $[0, \pi]$, *we have*

$$\cos x = \frac{1}{2} \iff x = \arccos\left(\frac{1}{2}\right) = \frac{\pi}{3}.$$

And since arctan *is the inverse function of* tan *on* $\left(-\frac{\pi}{2}, \frac{\pi}{2}\right)$, *we obtain*

$$\tan x = 1 \iff x = \arctan(1) = \frac{\pi}{4}.$$

Exercise 374. *Prove that for all* $x \in \mathbb{R}$,

$$\sin^2 x + \cos^2 x = 1,$$

and deduce from this the estimates

$$|\sin x| \leq 1 \qquad \textit{and} \qquad |\cos x| \leq 1.$$

Solution 375. *The point on the unit circle corresponding to the angle x has coordinates*

$$(\cos x, \sin x).$$

Since this point lies on the unit circle, it satisfies the equation

$$(\cos x)^2 + (\sin x)^2 = 1.$$

Thus, for all $x \in \mathbb{R}$,

$$\sin^2 x + \cos^2 x = 1.$$

Since both summands are nonnegative, it follows immediately that

$$\sin^2 x \leq 1 \qquad \text{and} \qquad \cos^2 x \leq 1.$$

Hence we obtain

$$|\sin x| \leq 1 \qquad \text{and} \qquad |\cos x| \leq 1.$$

Chapter 12

Differentiation

After the fundamental concepts of sequences, limits, and continuous functions, we now turn to one of the central topics of analysis: *differentiation.* While the notion of a limit first describes what happens to a sequence or a function near a point, differentiation is concerned with a more refined question: *How rapidly does a function change at a point?*

The idea of the derivative is closely connected with the geometric notion of a tangent. Instead of considering merely a secant line joining two points of the graph of a function, we consider the limiting case in which these two points approach each other more and more closely. The resulting limit measures the instantaneous rate of change of the function at the point in question. Thus the derivative provides not only a local measure of change, but also a crucial approach to the linear approximation of functions.

In this chapter, we develop the concept of differentiation systematically. First, we specify precisely when a function is differentiable and how the corresponding difference quotient is defined. We then examine the most important rules of differentiation, elementary examples, and fundamental consequences of differentiability. In particular, it will become clear that differentiability is a stronger property than continuity.

12.1 Differentiability of Functions

Let $D \subset \mathbb{R}$ and let

$$f : D \to \mathbb{R}$$

be a function. We now introduce precisely what it means for f to be differentiable at a point $x \in D$.

Definition 376. *The function f is called* differentiable *at a point $x \in D$ if the limit*

$$f'(x) := \lim_{\substack{a \to x \\ a \in D \setminus \{x\}}} \frac{f(a) - f(x)}{a - x}$$

exists.

The limit $f'(x)$ is called the difference quotient in the limit *or the* derivative *of f at the point x.*

For this limit to exist, the values of

$$\frac{f(a) - f(x)}{a - x}$$

must approach one and the same limiting value for all points $a \in D$, $a \neq x$, that tend to x. In particular, it is important that there actually exist sequences in $D \setminus \{x\}$ that converge to x. Otherwise the limiting process has no meaningful content; in such a case one usually does not speak of differentiability in the ordinary sense.

A particularly useful and often more convenient notation is obtained by the substitution

$$h := a - x.$$

Then $a = x + h$, and the definition above can equivalently be written as

$$f'(x) = \lim_{h \to 0} \frac{f(x+h) - f(x)}{h}.$$

Here only those sequences $(h_n)_{n \in \mathbb{N}}$ are admissible for which

$$\lim_{n \to \infty} h_n = 0, \qquad h_n \neq 0 \text{ for all } n, \qquad x + h_n \in D \text{ for all } n$$

hold.

Definition 377. *The function f is called* differentiable *on D if it is differentiable at every point $x \in D$.*

Thus the derivative $f'(x)$ describes the local rate of change of f at the point x. It is therefore a central tool for investigating the behavior of functions near a point. Later we shall see that it can be used not only to determine tangent lines, but also to analyze monotonicity, extrema, and many other properties.

12.2 Geometric Interpretation

The derivative can be interpreted not only analytically, but also geometrically. Let

$$P = (x, f(x)) \qquad \text{and} \qquad Q = (a, f(a))$$

be two points on the graph of f. Then the difference quotient

$$\frac{f(a) - f(x)}{a - x}$$

is the slope of the secant line through the two points P and Q.

Geometrically, this means that the numerator $f(a) - f(x)$ describes the *vertical change*, while the denominator $a - x$ describes the *horizontal change*. The quotient is therefore precisely the ratio of vertical change to horizontal distance, that is, the slope of the line connecting the two points.

If we now let $a \to x$, then the point Q moves closer and closer to P. In the limiting case, the secant line becomes a tangent line. If the limit of the difference quotient exists, then it is the slope of this tangent line. Hence

$$f'(x) = \text{slope of the tangent at the point } (x, f(x)).$$

For the derivative, one also uses the Leibniz notation

$$f'(x) = \frac{df}{dx}(x),$$

often written briefly as

$$\frac{df}{dx}.$$

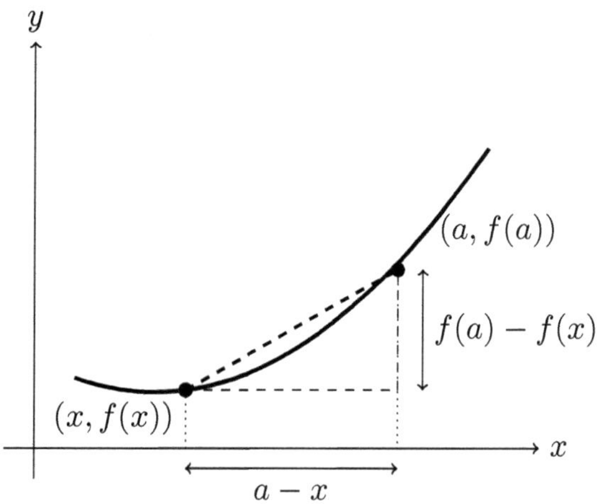

Figure 12.1: The slope of the secant line through the points $(x, f(x))$ and $(a, f(a))$ is the difference quotient $\dfrac{f(a) - f(x)}{a - x}$.

If we now let a tend to x, then the secant line approaches a limiting position. If the limit of the difference quotient exists, then we obtain the tangent to the graph at the point $(x, f(x))$. Its slope is the derivative $f'(x)$.

12.3 Examples for Computing Derivatives

Example 378 (Constant function). *Let*

$$f(x) = c$$

with a constant $c \in \mathbb{R}$. Then, for every $x \in D$,

$$f'(x) = \lim_{a \to x} \frac{f(a) - f(x)}{a - x} = \lim_{a \to x} \frac{c - c}{a - x} = \lim_{a \to x} 0 = 0.$$

Thus a constant function has derivative 0 *everywhere.*

Example 379 (Linear function). *Let*

$$f(x) = mx + b$$

with $m, b \in \mathbb{R}$. It follows that

$$f'(x) = \lim_{a \to x} \frac{f(a) - f(x)}{a - x} = \lim_{a \to x} \frac{(ma + b) - (mx + b)}{a - x}.$$

Simplifying gives

$$f'(x) = \lim_{a \to x} \frac{m(a - x)}{a - x} = \lim_{a \to x} m = m.$$

Thus the derivative of a linear function is constant and equals its slope.

Example 380 (Quadratic function). *Let*

$$f(x) = x^2.$$

Then

$$f'(x) = \lim_{a \to x} \frac{a^2 - x^2}{a - x}.$$

Using the binomial formula, or equivalently the factorization $a^2 - x^2 = (a - x)(a + x)$, *we obtain*

$$f'(x) = \lim_{a \to x} \frac{(a - x)(a + x)}{a - x} = \lim_{a \to x} (a + x) = 2x.$$

Hence the derivative of x^2 *is given by*

$$\frac{d}{dx}(x^2) = 2x.$$

Example 381 (Cubic function). *Let*

$$f(x) = x^3.$$

Then

$$f'(x) = \lim_{a \to x} \frac{a^3 - x^3}{a - x}.$$

Since

$$a^3 - x^3 = (a - x)(a^2 + ax + x^2),$$

it follows that

$$f'(x) = \lim_{a \to x} (a^2 + ax + x^2) = x^2 + x^2 + x^2 = 3x^2.$$

Therefore,

$$\frac{d}{dx}(x^3) = 3x^2.$$

Example 382 (Reciprocal function). *Let*

$$f(x) = \frac{1}{x} \qquad (x \neq 0).$$

Then, for $x \neq 0$, we compute:

$$f'(x) = \lim_{a \to x} \frac{\frac{1}{a} - \frac{1}{x}}{a - x} = \lim_{a \to x} \frac{x - a}{ax(a - x)}.$$

Since $x - a = -(a - x)$, the fraction simplifies to

$$f'(x) = \lim_{a \to x} \left(-\frac{1}{ax} \right).$$

Letting $a \to x$, we therefore obtain

$$f'(x) = -\frac{1}{x^2}.$$

Remark 383. *The examples illustrate the general procedure: one substitutes the function into the difference quotient, simplifies algebraically, and then takes the limit.*

Example 384 (Sine function). *Let*

$$f(x) = \sin(x).$$

Then

$$f'(x) = \lim_{a \to x} \frac{\sin(a) - \sin(x)}{a - x}.$$

We set

$$a = x + h \qquad \text{with } h \to 0.$$

It follows that

$$f'(x) = \lim_{h \to 0} \frac{\sin(x+h) - \sin(x)}{h}.$$

Using the addition formula for the sine, we obtain

$$\sin(x+h) = \sin(x)\cos(h) + \cos(x)\sin(h).$$

Hence

$$\frac{\sin(x+h) - \sin(x)}{h} = \sin(x)\frac{\cos(h)-1}{h} + \cos(x)\frac{\sin(h)}{h}.$$

Therefore,

$$f'(x) = \sin(x)\lim_{h \to 0}\frac{\cos(h)-1}{h} + \cos(x)\lim_{h \to 0}\frac{\sin(h)}{h}.$$

Using the limits

$$\lim_{h \to 0}\frac{\sin(h)}{h} = 1 \qquad \textit{and} \qquad \lim_{h \to 0}\frac{\cos(h)-1}{h} = 0$$

we obtain

$$f'(x) = \cos(x).$$

Thus the derivative of the sine function is the cosine function.

Example 385 (Cosine function). *Let*

$$f(x) = \cos(x).$$

Then

$$f'(x) = \lim_{a \to x}\frac{\cos(a) - \cos(x)}{a - x}.$$

Using the substitution

$$a = x + h, \qquad h \to 0,$$

we obtain

$$f'(x) = \lim_{h\to 0} \frac{\cos(x+h) - \cos(x)}{h}.$$

By the addition formula for the cosine, one has

$$\cos(x+h) = \cos(x)\cos(h) - \sin(x)\sin(h).$$

Thus

$$\frac{\cos(x+h) - \cos(x)}{h} = \cos(x)\frac{\cos(h) - 1}{h} - \sin(x)\frac{\sin(h)}{h}.$$

It follows that

$$f'(x) = \cos(x)\lim_{h\to 0}\frac{\cos(h) - 1}{h} - \sin(x)\lim_{h\to 0}\frac{\sin(h)}{h}.$$

Again, using

$$\lim_{h\to 0}\frac{\cos(h) - 1}{h} = 0 \qquad \textit{and} \qquad \lim_{h\to 0}\frac{\sin(h)}{h} = 1$$

we obtain

$$f'(x) = -\sin(x).$$

Thus, the derivative of the cosine function is minus sine.

Example 386 (Exponential function). *Let*

$$f(x) = e^x.$$

Then

$$f'(x) = \lim_{a \to x} \frac{e^a - e^x}{a - x}.$$

With the substitution

$$a = x + h, \qquad h \to 0,$$

we obtain

$$f'(x) = \lim_{h \to 0} \frac{e^{x+h} - e^x}{h}.$$

Since

$$e^{x+h} = e^x e^h,$$

it follows that

$$f'(x) = \lim_{h \to 0} e^x \frac{e^h - 1}{h} = e^x \lim_{h \to 0} \frac{e^h - 1}{h}.$$

Using the fundamental limit

$$\lim_{h \to 0} \frac{e^h - 1}{h} = 1$$

we get

$$f'(x) = e^x.$$

Thus, the exponential function is its own derivative.

Theorem 387 (Local linear approximation). *Let $D \subset \mathbb{R}$, and let $a \in D$ be an accumulation point of D, that is, a point for which there exists a sequence $(x_n)_{n \in \mathbb{N}}$ such that*

$$x_n \in D \setminus \{a\} \quad \text{for all } n \qquad \text{and} \qquad \lim_{n \to \infty} x_n = a$$

holds.

Then a function $f : D \to \mathbb{R}$ is differentiable at a if and only if there exists a constant $c \in \mathbb{R}$ and a function $\phi : D \to \mathbb{R}$ such that

$$f(x) = f(a) + c(x - a) + \phi(x)$$

for all $x \in D$, where

$$\lim_{\substack{x \to a \\ x \neq a}} \frac{\phi(x)}{x - a} = 0$$

holds.

In this case, necessarily

$$c = f'(a).$$

Proof. We first prove the implication "$\Rightarrow$".

Assume that f is differentiable at a. Then the limit

$$f'(a) = \lim_{\substack{x \to a \\ x \neq a}} \frac{f(x) - f(a)}{x - a}$$

exists. We now set

$$c := f'(a)$$

and define, for $x \in D$, the function

$$\phi(x) := f(x) - f(a) - c(x-a).$$

Then, for every $x \in D$, we have

$$f(x) = f(a) + c(x-a) + \phi(x).$$

Moreover, for $x \neq a$,

$$\frac{\phi(x)}{x-a} = \frac{f(x) - f(a) - c(x-a)}{x-a} = \frac{f(x)-f(a)}{x-a} - c.$$

Taking the limit as $x \to a$, we obtain

$$\lim_{\substack{x \to a \\ x \neq a}} \frac{\phi(x)}{x-a} = \lim_{\substack{x \to a \\ x \neq a}} \frac{f(x)-f(a)}{x-a} - c = f'(a) - c = 0.$$

Thus the desired representation has been found.

We now prove the implication "$\Leftarrow$".

Assume that there is a constant $c \in \mathbb{R}$ and a function $\phi : D \to \mathbb{R}$ such that

$$f(x) = f(a) + c(x-a) + \phi(x)$$

and

$$\lim_{\substack{x \to a \\ x \neq a}} \frac{\phi(x)}{x-a} = 0.$$

Then, for $x \neq a$, it follows that

$$\frac{f(x)-f(a)}{x-a} = c + \frac{\phi(x)}{x-a}.$$

Since the second term converges to 0, the limit of the difference quotient exists, namely

$$\lim_{\substack{x \to a \\ x \neq a}} \frac{f(x) - f(a)}{x - a} = c.$$

Hence f is differentiable at a, and

$$f'(a) = c.$$

This proves the equivalence. □

Differentiability at a therefore means precisely that, near a, the function f can be approximated by the affine function

$$x \mapsto f(a) + f'(a)(x - a)$$

up to an error term that vanishes in comparison with $x - a$.

Corollary 388 (Differentiability implies continuity). *Let $D \subset \mathbb{R}$ and let $a \in D$ be an accumulation point of D. If a function $f : D \to \mathbb{R}$ is differentiable at the point a, then f is continuous at a.*

Proof. By the preceding theorem, there exists a constant $c \in \mathbb{R}$ and a function $\phi : D \to \mathbb{R}$ such that

$$f(x) = f(a) + c(x - a) + \phi(x)$$

for all $x \in D$ and

$$\lim_{\substack{x \to a \\ x \neq a}} \frac{\phi(x)}{x - a} = 0.$$

We now consider the limiting process $x \to a$.

Clearly,

$$\lim_{x \to a} c(x-a) = 0.$$

Moreover, from

$$\frac{\phi(x)}{x-a} \to 0 \quad \text{and} \quad x-a \to 0$$

it follows that also

$$\phi(x) = (x-a) \cdot \frac{\phi(x)}{x-a} \to 0.$$

Thus we obtain

$$\lim_{x \to a} f(x) = \lim_{x \to a} (f(a)+c(x-a)+\phi(x)) = f(a)+0+0 = f(a).$$

Hence f is continuous at a. □

In what follows, we show that differentiability is compatible with the elementary algebraic operations. In particular, the derivative of linear combinations of functions, as well as of products and quotients of differentiable functions, can again be expressed solely in terms of their derivatives. These rules are essential, since they reduce the practical computation of derivatives of more complex functions to already known cases.

Theorem 389 (Rules for derivatives). *Let $D \subset \mathbb{R}$ and $x \in D$. Moreover, let $f, g : D \to \mathbb{R}$ be differentiable at x, and let $\lambda \in \mathbb{R}$.*

Then $f+g$, λf, and fg are also differentiable at x, and the following rules hold:

$$(f+g)'(x) = f'(x) + g'(x), \qquad (\lambda f)'(x) = \lambda f'(x),$$

and

$$(fg)'(x) = f'(x)g(x) + f(x)g'(x).$$

If, in addition, $g(a) \neq 0$ for all $a \in D$, then $\frac{f}{g}$ is also differentiable at x, and

$$\left(\frac{f}{g}\right)'(x) = \frac{f'(x)g(x) - f(x)g'(x)}{g(x)^2}.$$

Proof. We begin with linearity.

From the differentiability of f and g at x, we obtain the limits

$$f'(x) = \lim_{\substack{a \to x \\ a \neq x}} \frac{f(a) - f(x)}{a - x}, \qquad g'(x) = \lim_{\substack{a \to x \\ a \neq x}} \frac{g(a) - g(x)}{a - x}.$$

For the sum, we get

$$\frac{(f+g)(a) - (f+g)(x)}{a - x} = \frac{f(a) - f(x)}{a - x} + \frac{g(a) - g(x)}{a - x}.$$

Since limits are additive, it follows that

$$(f+g)'(x) = f'(x) + g'(x).$$

Similarly, for λf we have

$$\frac{(\lambda f)(a) - (\lambda f)(x)}{a - x} = \lambda \frac{f(a) - f(x)}{a - x},$$

hence
$$(\lambda f)'(x) = \lambda f'(x).$$

We now turn to the product rule. For $a \neq x$, we write

$$f(a)g(a)-f(x)g(x) = f(a)(g(a)-g(x))+g(x)(f(a)-f(x)).$$

Thus

$$\frac{f(a)g(a) - f(x)g(x)}{a - x} = f(a)\frac{g(a) - g(x)}{a - x}+g(x)\frac{f(a) - f(x)}{a - x}.$$

Letting $a \to x$, we have $f(a) \to f(x)$ because f is differentiable and hence, in particular, continuous at x. Moreover, the two derivative limits of f and g exist. Therefore,

$$(fg)'(x) = f(x)g'(x) + g(x)f'(x).$$

For the quotient rule, we first consider the reciprocal function
$$r(a) := \frac{1}{g(a)}.$$

Since $g(a) \neq 0$ for all $a \in D$, the function r is well-defined. From
$$g(a)\,r(a) = 1$$
we obtain, by differentiating with the product rule,

$$g'(x)r(x) + g(x)r'(x) = 0.$$

Since $r(x) = \frac{1}{g(x)}$, we obtain

$$r'(x) = -\frac{g'(x)}{g(x)^2}.$$

Now we write
$$\frac{f}{g} = f \cdot r.$$
Hence, by the product rule,
$$\begin{aligned}\left(\frac{f}{g}\right)'(x) &= f'(x)r(x) + f(x)r'(x)\\ &= \frac{f'(x)}{g(x)} - f(x)\frac{g'(x)}{g(x)^2}\\ &= \frac{f'(x)g(x) - f(x)g'(x)}{g(x)^2}\end{aligned}$$
This proves all the rules. □

Example 390 (Derivative using differentiation rules). *We consider the function*
$$f(x) = (x^2)\sin(x).$$

To compute the derivative, we use the product rule:
$$(fg)'(x) = f'(x)g(x) + f(x)g'(x).$$

Set
$$f_1(x) = x^2, \qquad f_2(x) = \sin(x).$$
Then
$$f_1'(x) = 2x, \qquad f_2'(x) = \cos(x).$$
It follows that
$$f'(x) = 2x\sin(x) + x^2\cos(x).$$

Thus,

$$\frac{d}{dx}(x^2 \sin(x)) = 2x \sin(x) + x^2 \cos(x).$$

Many important functions have an inverse function. If such a function is differentiable, it is natural to expect that the derivative of the inverse function can also be determined from the derivative of the original function. Indeed, under suitable assumptions, a remarkably simple relation holds: the derivative of the inverse function is the reciprocal of the derivative of the original function.

Theorem 391 (Derivative of the inverse function)**.** *Let $D \subset \mathbb{R}$ be an interval and let*

$$f : D \to \mathbb{R}$$

be a continuous, strictly monotone function. Furthermore, let

$$E := f(D) \quad \text{and} \quad \phi := f^{-1} : E \to D$$

be the inverse function of f.

If f is differentiable at a point $x \in D$ and

$$f'(x) \neq 0,$$

then ϕ is differentiable at the point

$$y := f(x)$$

and

$$\phi'(y) = \frac{1}{f'(x)} = \frac{1}{f'(\phi(y))}.$$

Proof. We consider the difference quotient of the inverse function at the point $y = f(x)$. Let $a \in E$ with $a \neq y$. Since $a \in E = f(D)$, we can write

$$a = f(b)$$

for a uniquely determined $b \in D$. Since $\phi = f^{-1}$, we have $b = \phi(a)$. Moreover,

$$\phi(y) = \phi(f(x)) = x.$$

Thus the difference quotient becomes

$$\frac{\phi(a) - \phi(y)}{a - y} = \frac{b - x}{f(b) - f(x)}.$$

For $b \neq x$, this can be rewritten as

$$\frac{b - x}{f(b) - f(x)} = \frac{1}{\frac{f(b) - f(x)}{b - x}}.$$

We now let $a \to y$. Since $a = f(b)$ and f is continuous and strictly monotone, it follows that $b \to x$ as well. Since f is differentiable at x, we have

$$\lim_{b \to x} \frac{f(b) - f(x)}{b - x} = f'(x).$$

Because of the assumption $f'(x) \neq 0$, we may take reciprocals, and hence obtain

$$\lim_{a \to y} \frac{\phi(a) - \phi(y)}{a - y} = \lim_{b \to x} \frac{1}{\frac{f(b) - f(x)}{b - x}} = \frac{1}{f'(x)}.$$

Therefore ϕ is differentiable at y, and

$$\phi'(y) = \frac{1}{f'(x)}.$$

Since $y = f(x)$ and therefore $x = \phi(y)$, it also follows that

$$\phi'(y) = \frac{1}{f'(\phi(y))}.$$

□

Example 392 (Derivative of the natural logarithm). *We consider the exponential function*

$$f(x) = e^x.$$

It is continuous, strictly increasing, and differentiable on all of $\mathbb{R}$. *Moreover,*

$$f'(x) = e^x > 0$$

for all $x \in \mathbb{R}$.

The inverse function of f *is the natural logarithm*

$$\ln : (0, \infty) \to \mathbb{R}.$$

Now let $y > 0$. *Since* $\ln(y)$ *is precisely the number for which*

$$e^{\ln(y)} = y$$

holds, we can apply the theorem on the derivative of the inverse function. We obtain

$$(\ln)'(y) = \frac{1}{f'(\ln(y))} = \frac{1}{e^{\ln(y)}}.$$

Since $e^{\ln(y)} = y$, *it follows that*

$$(\ln)'(y) = \frac{1}{y}.$$

Thus, for all $y > 0$,

$$\boxed{\frac{d}{dy}\ln(y) = \frac{1}{y}.}$$

The differentiation rules discussed so far allow us to differentiate sums, products, quotients, and inverse functions. For many important functions, however, this is not sufficient, since they arise as compositions of simpler functions. In order to differentiate such functions as well, we need the chain rule.

Theorem 393 (Chain rule). *Let* $D, E \subset \mathbb{R}$,

$$f : D \to \mathbb{R}, \qquad g : E \to \mathbb{R},$$

and suppose that $f(D) \subset E$. *If* f *is differentiable at the point* $x \in D$ *and* g *is differentiable at the point*

$$y := f(x) \in E$$

then the composite function

$$g \circ f : D \to \mathbb{R}$$

is also differentiable at x, *and*

$$(g \circ f)'(x) = g'(f(x))\, f'(x).$$

Proof. We first rewrite the difference quotient of $g \circ f$ in such a way that the differentiability of g becomes visible.

For this purpose, consider the function

$$\gamma : E \to \mathbb{R}$$

defined by

$$\gamma(t) := \begin{cases} \dfrac{g(t) - g(y)}{t - y}, & t \neq y, \\ g'(y), & t = y. \end{cases}$$

Since g is differentiable at y, we have

$$\lim_{t \to y} \gamma(t) = g'(y).$$

In particular, γ is continuous at y.

For $a \in D$ with $a \neq x$, we now have

$$g(f(a)) - g(f(x)) = \gamma(f(a))\,(f(a) - f(x)).$$

Thus the difference quotient of $g \circ f$ becomes

$$\frac{(g \circ f)(a) - (g \circ f)(x)}{a - x} = \gamma(f(a)) \cdot \frac{f(a) - f(x)}{a - x}.$$

Now let $a \to x$. Since f is differentiable at x, it follows in particular that

$$f(a) \to f(x) = y.$$

By the continuity of γ at y, we therefore obtain

$$\gamma(f(a)) \to \gamma(y) = g'(y) = g'(f(x)).$$

Moreover,

$$\frac{f(a) - f(x)}{a - x} \to f'(x).$$

Taking limits in the product representation therefore gives

$$(g \circ f)'(x) = g'(f(x))\, f'(x).$$

This proves the chain rule. □

Example 394 (Derivative of a composite function). *Consider the function*

$$f(x) = \ln(1 + x^2).$$

This function is not directly an elementary standard function, but arises as the composition of two simpler functions.

To see this, set

$$u(x) = 1 + x^2 \qquad \textit{and} \qquad v(x) = \ln(x).$$

Then

$$f(x) = (v \circ u)(x) = v(u(x)).$$

By the chain rule, we obtain

$$f'(x) = v'(u(x)) \cdot u'(x).$$

Since

$$v'(x) = \frac{1}{x} \qquad \textit{and} \qquad u'(x) = 2x,$$

it follows that

$$f'(x) = \frac{1}{1 + x^2} \cdot 2x.$$

Hence

$$\frac{d}{dx}\ln(1+x^2) = \frac{2x}{1+x^2}.$$

This example is particularly useful because it shows the chain rule together with a typical function from analysis.

The differentiation rules developed so far can be combined with one another and thereby make it possible to differentiate further types of functions. In particular, a quotient can be regarded as the product of one function with the reciprocal of another. With the help of the chain rule, one can first determine the derivative of a reciprocal function. Together with the product rule, this immediately yields the quotient rule.

Theorem 395 (Quotient rule). *Let $D \subset \mathbb{R}$, and let $g : D \to \mathbb{R}$ be differentiable at a point $x \in D$, where*

$$g(x) \neq 0.$$

Then the function

$$\frac{1}{g} : D \to \mathbb{R}, \qquad \left(\frac{1}{g}\right)(x) = \frac{1}{g(x)},$$

is also differentiable at x. Moreover,

$$\left(\frac{1}{g}\right)'(x) = -\frac{g'(x)}{g(x)^2}.$$

If, in addition, $f : D \to \mathbb{R}$ is differentiable at x, then $\frac{f}{g}$ is also differentiable at x, and

$$\left(\frac{f}{g}\right)'(x) = \frac{f'(x)g(x) - f(x)g'(x)}{g(x)^2}.$$

Proof. We begin with the function

$$r(t) = \frac{1}{t} \qquad (t \neq 0).$$

Its derivative is

$$r'(t) = -\frac{1}{t^2}.$$

Since $g(x) \neq 0$ and g is differentiable at x, we may apply the chain rule to the composition

$$\frac{1}{g} = r \circ g$$

and obtain

$$\left(\frac{1}{g}\right)'(x) = r'(g(x)) \cdot g'(x) = -\frac{1}{g(x)^2}\, g'(x).$$

Thus

$$\left(\frac{1}{g}\right)'(x) = -\frac{g'(x)}{g(x)^2}.$$

Now we write

$$\frac{f}{g} = f \cdot \frac{1}{g}.$$

By the product rule, we obtain

$$\left(\frac{f}{g}\right)'(x) = f'(x)\frac{1}{g(x)} + f(x)\left(\frac{1}{g}\right)'(x).$$

Substituting the derivative of $\frac{1}{g}$ computed above gives

$$\left(\frac{f}{g}\right)'(x) = \frac{f'(x)}{g(x)} - f(x)\frac{g'(x)}{g(x)^2}.$$

Putting the terms over a common denominator yields

$$\left(\frac{f}{g}\right)'(x) = \frac{f'(x)g(x) - f(x)g'(x)}{g(x)^2}.$$

This proves the quotient rule. □

Definition 396 (Higher derivatives). *Let $D \subset \mathbb{R}$ and let*

$$f : D \to \mathbb{R}$$

be a function. We first define the zeroth derivative *of f by*

$$f^{(0)} := f.$$

If $f^{(k-1)}$ is differentiable at a point $x \in D$, then f is said to be k times differentiable *at x, and we define the* k-th derivative *of f at x by*

$$f^{(k)}(x) := (f^{(k-1)})'(x).$$

Thus the k-th derivative is obtained step by step from the $(k-1)$-st derivative.

In particular, this means:

$$f^{(1)} = f', \qquad f^{(2)} = (f')', \qquad f^{(3)} = ((f')')', \quad \textit{etc.}$$

If f is k times differentiable at every point of D, then f is said to be k times differentiable on *D. In this case, $f^{(k)}$ is again itself a function on D.*

If f is k times differentiable on D and the k-th derivative $f^{(k)}$ is continuous on D, then f is called k

times continuously differentiable *on D. This is also written as*

$$f \in C^k(D).$$

A function is called infinitely differentiable *on D if it is k times differentiable on D for every $k \in \mathbb{N}$.*

Remark 397. *The definition is inductive in nature: from the first derivative one obtains the second, from the second the third, and so on. In this way, an entire sequence of derivatives arises,*

$$f,\ f',\ f'',\ f^{(3)},\ \dots$$

which describes the local behavior of a function with increasing precision.

Derivatives therefore provide not only a tool for describing local rates of change, but also the foundation for studying extrema and mean-value behavior. In the following chapter, we therefore discuss local extrema and the mean value theorem.

12.4 Exercises and Solutions

Exercise 398. *Determine the derivative of the function*

$$f(x) = x^3 - 2x^2 + 5x - 1.$$

Solution 399. *We differentiate each term separately. Using the rules of differentiation, we obtain*

$$(x^3)' = 3x^2, \qquad (-2x^2)' = -4x,$$
$$(5x)' = 5, \qquad (-1)' = 0.$$

Therefore,

$$f'(x) = 3x^2 - 4x + 5.$$

Exercise 400. *Determine the derivative of the function*

$$f(x) = (x^2 + 1)\sin(x).$$

Solution 401. *This is a product. We set*

$$u(x) = x^2 + 1, \qquad v(x) = \sin(x).$$

Then

$$u'(x) = 2x, \qquad v'(x) = \cos(x).$$

By the product rule, it follows that

$$f'(x) = u'(x)v(x)+u(x)v'(x) = 2x\sin(x)+(x^2+1)\cos(x).$$

Exercise 402. *Determine the derivative of the function*

$$f(x) = e^{3x-1}.$$

Solution 403. *This function is a composition. We write*

$$u(x) = 3x - 1, \qquad v(x) = e^x,$$

so that

$$f(x) = v(u(x)).$$

Then

$$u'(x) = 3, \qquad v'(x) = e^x.$$

Using the chain rule, we obtain

$$f'(x) = v'(u(x)) \cdot u'(x) = e^{3x-1} \cdot 3.$$

Hence,

$$f'(x) = 3e^{3x-1}.$$

Exercise 404. *Determine the derivative of the function*

$$f(x) = \ln(1 + x^2).$$

Solution 405. *Here again we have a composition. We set*

$$u(x) = 1 + x^2, \qquad v(x) = \ln(x).$$

Then

$$f(x) = v(u(x)).$$

We have

$$u'(x) = 2x, \qquad v'(x) = \frac{1}{x}.$$

Using the chain rule, it follows that

$$f'(x) = v'(u(x)) \cdot u'(x) = \frac{1}{1 + x^2} \cdot 2x.$$

Hence we obtain

$$f'(x) = \frac{2x}{1 + x^2}.$$

Exercise 406. *Determine the derivative of the function*

$$f(x) = \frac{x^2 + 1}{x - 1} \qquad (x \neq 1).$$

Solution 407. *We use the quotient rule with*

$$u(x) = x^2 + 1, \qquad v(x) = x - 1.$$

Then

$$u'(x) = 2x, \qquad v'(x) = 1.$$

Therefore,

$$f'(x) = \frac{u'(x)v(x) - u(x)v'(x)}{v(x)^2} = \frac{2x(x-1) - (x^2+1)}{(x-1)^2}.$$

Expanding gives

$$f'(x) = \frac{2x^2 - 2x - x^2 - 1}{(x-1)^2} = \frac{x^2 - 2x - 1}{(x-1)^2}.$$

Exercise 408. *Determine the equation of the tangent line to the graph of the function*

$$f(x) = x^2$$

at the point $x = 1$.

Solution 409. *First, we compute the derivative:*

$$f'(x) = 2x.$$

Thus the slope of the tangent line at the point $x = 1$ *is*

$$f'(1) = 2.$$

Moreover,

$$f(1) = 1.$$

Therefore, the tangent line to the graph at the point $(1, 1)$ *has the equation*

$$y = f(1) + f'(1)(x - 1) = 1 + 2(x - 1).$$

After expanding, we obtain

$$y = 2x - 1.$$

Exercise 410. *Show that the function*

$$f(x) = x^2 \sin(x)$$

is differentiable, and compute its derivative.

Solution 411. *The function is a product of two differentiable functions, namely*

$$u(x) = x^2 \qquad \textit{and} \qquad v(x) = \sin(x).$$

Therefore, f is differentiable. Moreover,

$$u'(x) = 2x, \qquad v'(x) = \cos(x).$$

By the product rule, it follows that

$$f'(x) = u'(x)v(x) + u(x)v'(x) = 2x\sin(x) + x^2\cos(x).$$

Exercise 412. *Let*

$$f(x) = \ln(x).$$

Determine the derivative of f^{-1} and interpret the result.

Solution 413. *The inverse function of $\ln(x)$ is the exponential function e^x. Since*

$$(\ln)'(x) = \frac{1}{x}$$

holds for $x > 0$, the theorem on the derivative of the inverse function gives

$$(e^x)' = \frac{1}{(\ln)'(e^x)} = \frac{1}{1/e^x} = e^x.$$

Thus, the exponential function is its own derivative.

Chapter 13

Local Extrema and the Mean Value Theorem

After the introduction of the derivative, the natural question arises as to what information it actually provides about the behavior of a function. The answer is of central importance in analysis: derivatives can be used to identify local extrema, derive monotonicity properties, and estimate rates of growth. Thus differentiation becomes one of the most important tools of all, not only in pure mathematics, but also in fields of application such as optimization, physics, economics, and the natural sciences.

The basic idea is simple: if a function is particularly large or particularly small at a point, then in an immediate neighborhood of that point it can no longer continue to increase or decrease, respectively. In many cases, this leads to the derivative vanishing at that point. Conversely, a vanishing derivative is an important indication of an extremum, but it is not yet a guarantee of one. For precisely this reason, it is worthwhile first to define local extrema carefully and then to examine their relationship with the derivative.

In the course of this chapter, we shall see that the Mean Value Theorem extends this local point of view into a powerful global instrument. It connects function values at two points with the derivative on an entire interval and thereby forms the basis for numerous further results in analysis.

13.1 Local Extrema

Definition 414. *A function* $f :]a,b[\to \mathbb{R}$ *has a* local maximum *at* $x \in]a,b[$ *if there exists an* $\varepsilon > 0$ *such that*

$$f(x) \geq f(\varphi) \qquad \text{for all } \varphi \in]a,b[\text{ with } |x - \varphi| < \varepsilon.$$

Analogously, x *is called a* local minimum *if there exists an* $\varepsilon > 0$ *such that*

$$f(x) \leq f(\varphi) \qquad \text{for all } \varphi \in]a,b[\text{ with } |x - \varphi| < \varepsilon.$$

If the respective inequality is even strict for all $\varphi \neq x$ *with* $|x - \varphi| < \varepsilon$*, then one speaks of a* strict *local maximum or minimum, respectively.*

A local maximum and a local minimum are collectively called a local extremum.

Theorem 415. *Suppose that a function $f :]a, b[\to \mathbb{R}$ has a local extremum at the point $x \in]a, b[$ and is differentiable at x. Then*

$$f'(x) = 0.$$

Proof. We first consider the case in which f has a local maximum at x. Then there exists an $\varepsilon > 0$ such that

$$f(\varphi) \leq f(x) \qquad \text{for all } \varphi \in]x - \varepsilon, x + \varepsilon[.$$

For $h > 0$ with $x + h \in]x - \varepsilon, x + \varepsilon[$, it follows that

$$\frac{f(x+h) - f(x)}{h} \leq 0,$$

so the right-hand derivative is at most 0:

$$f'_+(x) \leq 0.$$

For $h < 0$, on the other hand, since $f(x + h) - f(x) \leq 0$ and $h < 0$, we have

$$\frac{f(x+h) - f(x)}{h} \geq 0,$$

so the left-hand derivative is at least 0:

$$f'_-(x) \geq 0.$$

Since f is differentiable at x, the two one-sided derivatives agree. Hence both

$$f'(x) \leq 0 \quad \text{and} \quad f'(x) \geq 0$$

hold, and therefore

$$f'(x) = 0.$$

In the case of a local minimum, the inequalities are reversed, and one likewise obtains $f'(x) = 0$. □

Remark 416. *The condition $f'(x) = 0$ is necessary, but not sufficient, for a local extremum. A simple example is the constant function $f(x) = 0$, whose derivative vanishes everywhere, although not every point is an extremum.*

If an extremum occurs at an endpoint of a closed interval, then in general $f'(x) = 0$ need not hold there. For example, the function

$$f : [0, 1] \to \mathbb{R}, \qquad f(x) = x,$$

has an absolute maximum at $x = 1$, but the ordinary derivative does not exist at this endpoint; the left-hand derivative, however, is equal to 1.

Theorem 417 (Rolle's theorem)**.** *Let $f : [a, b] \to \mathbb{R}$ be a continuous function with $a < b$ that is differentiable on the open interval $]a, b[$. Suppose in addition that*

$$f(a) = f(b),$$

then there exists a point $\phi \in]a, b[$ such that

$$f'(\phi) = 0.$$

In particular, between two zeros of a differentiable function there is always at least one zero of its derivative.

Proof. If f is constant, then $f'(x) = 0$ for all $x \in]a, b[$, and the statement is immediate.

We may therefore assume that f is not constant. Then there is a point $x_0 \in [a, b]$ such that

$$f(x_0) \neq f(a) = f(b).$$

Thus either $f(x_0) > f(a)$ or $f(x_0) < f(a)$. In the first case,

$$\max_{x \in [a,b]} f(x) > f(a) = f(b),$$

and in the second case,

$$\min_{x \in [a,b]} f(x) < f(a) = f(b).$$

By the extreme value theorem, the continuous function f attains both its absolute maximum and its absolute minimum on the closed interval $[a, b]$. Since the function values at the endpoints are equal, in the nonconstant case at least one of these two extrema must be attained at an interior point $\phi \in]a, b[$.

At this point, f has a local extremum. Since f is differentiable there, the preceding theorem implies

$$f'(\phi) = 0.$$

This proves the theorem. □

Remark 418. *Rolle's theorem describes a simple but fundamental connection between global and local information: if a function has the same value at both endpoints, then somewhere in between it must have at least one horizontal tangent.*

13.2 The Mean Value Theorem

Theorem 419 (Mean Value Theorem of Differential Calculus). *Let $f : [a, b] \to \mathbb{R}$ be a continuous function that is differentiable on the open interval $]a, b[$. Then there exists a point $\phi \in]a, b[$ such that*

$$f'(\phi) = \frac{f(b) - f(a)}{b - a}.$$

Proof. We consider the line passing through the two points $(a, f(a))$ and $(b, f(b))$. Its slope is

$$m = \frac{f(b) - f(a)}{b - a}.$$

Now we introduce the auxiliary function

$$g(x) = f(x) - (f(a) + m(x - a))$$

Then g is also continuous on $[a, b]$ and differentiable on $]a, b[$.

First, we have

$$g(a) = f(a) - (f(a) + m(a - a)) = 0$$

and similarly

$$\begin{aligned} g(b) &= f(b) - (f(a) + m(b - a)) \\ &= f(b) - (f(a) + f(b) - f(a)) = 0. \end{aligned}$$

Thus g has the same value at the endpoints of the interval. By Rolle's theorem, there therefore exists a point $\phi \in]a, b[$ such that

$$g'(\phi) = 0.$$

Since

$$g'(x) = f'(x) - m,$$

it follows that

$$0 = g'(\phi) = f'(\phi) - m.$$

Hence

$$f'(\phi) = m = \frac{f(b) - f(a)}{b - a}.$$

This proves the assertion. □

Remark 420. *The mean value theorem has a very intuitive geometric meaning: The secant line through the points $(a, f(a))$ and $(b, f(b))$ has slope*

$$\frac{f(b) - f(a)}{b - a}.$$

The mean value theorem states that between a and b there is at least one point ϕ at which the tangent to the graph has the same slope as this secant line.

In other words, somewhere between the two endpoints, the instantaneous rate of change is equal to the average rate of change over the whole interval.

Corollary 421 (Estimate by the derivative)**.** *Let $f : [a, b] \to \mathbb{R}$ be continuous and differentiable on $]a, b[$. Suppose furthermore that*

$$m \leq f'(\phi) \leq M \qquad \textit{for all } \phi \in]a, b[$$

for two constant numbers $m, M \in \mathbb{R}$. Then, for all $x, y \in [a, b]$ with $x \leq y$, the estimate

$$m(y - x) \leq f(y) - f(x) \leq M(y - x).$$

holds.

Proof. Let $x, y \in [a, b]$ with $x < y$. Then we can apply the mean value theorem to the interval $[x, y]$. Hence there exists a point $\phi \in]x, y[$ such that

$$f'(\phi) = \frac{f(y) - f(x)}{y - x}.$$

By assumption,

$$m \leq f'(\phi) \leq M.$$

Multiplication by the positive number $y - x$ gives

$$m(y - x) \leq f(y) - f(x) \leq M(y - x).$$

For $x = y$, the assertion is trivial, since then both sides are equal to 0. □

Corollary 422 (Constancy for vanishing derivative). *Let $f : [a, b] \to \mathbb{R}$ be continuous and differentiable on $]a, b[$. If*

$$f'(x) = 0 \qquad \textit{for all } x \in]a, b[,$$

then f is constant on $[a, b]$.

Proof. This is a special case of the preceding corollary with $m = M = 0$. Then, for all $x, y \in [a, b]$ with $x \leq y$, it follows that

$$0 \leq f(y) - f(x) \leq 0.$$

Thus $f(y) - f(x) = 0$, and hence

$$f(y) = f(x)$$

for all $x, y \in [a, b]$. Therefore f is constant. □

Remark 423. *The first estimate is a precise quantitative form of the idea that the derivative controls the growth behavior of a function. If the derivative is bounded above everywhere by M, then the function cannot grow on an interval faster than a straight line with slope M. If it is additionally bounded below by m, then a corresponding minimum growth is also guaranteed.*

13.3 Monotonicity

The derivative describes not only the local behavior of a function, but also allows conclusions about its global growth behavior. If the derivative is nonnegative everywhere, then the function cannot decrease; if it is positive everywhere, then the function is even strictly increasing. Conversely, monotonicity forces the derivative to have the corresponding sign. The mean value theorem provides the central link between these facts.

Theorem 424 (Monotonicity criterion). *Let $f : [a,b] \to \mathbb{R}$ be a continuous function that is differentiable on the open interval $]a,b[$.*

1. *If*
$$f'(x) \geq 0 \qquad \text{for all } x \in]a,b[,$$
then f is increasing on $[a,b]$. If even
$$f'(x) > 0 \qquad \text{for all } x \in]a,b[,$$
then f is strictly increasing on $[a,b]$. Correspondingly, if
$$f'(x) \leq 0 \qquad \text{for all } x \in]a,b[$$
then f is decreasing; and if
$$f'(x) < 0 \qquad \text{for all } x \in]a,b[$$
then f is strictly decreasing.

2. *If f is increasing, then*
$$f'(x) \geq 0 \qquad \text{for all } x \in]a,b[.$$
If f is decreasing, then
$$f'(x) \leq 0 \qquad \text{for all } x \in]a,b[.$$

Proof. We first prove the first assertion in the case
$$f'(x) \geq 0 \qquad \text{for all } x \in]a,b[.$$

Let $x, y \in [a, b]$ with $x < y$. Then f is continuous on $[x, y]$ and differentiable on $]x, y[$. By the mean value theorem, there exists a point $\phi \in]x, y[$ such that

$$f'(\phi) = \frac{f(y) - f(x)}{y - x}.$$

Since, by assumption, $f'(\phi) \geq 0$ and also $y - x > 0$, it follows that

$$f(y) - f(x) \geq 0.$$

Thus

$$f(x) \leq f(y),$$

and therefore f is increasing.

If even $f'(x) > 0$ for all $x \in]a, b[$, then the same argument gives

$$f(y) - f(x) > 0 \qquad \text{for all } x < y,$$

so f is strictly increasing.

The case

$$f'(x) \leq 0 \qquad \text{for all } x \in]a, b[$$

is treated analogously: the mean value theorem gives

$$\frac{f(y) - f(x)}{y - x} = f'(\phi) \leq 0$$

for $x < y$, hence

$$f(y) - f(x) \leq 0.$$

Therefore f is decreasing; in the case $f'(x) < 0$, one obtains strict decrease.

For the second part, now suppose that f is increasing. Fix $x \in]a, b[$. Then, for all sufficiently small $h > 0$,

$$f(x+h) - f(x) \geq 0,$$

and hence

$$\frac{f(x+h) - f(x)}{h} \geq 0.$$

Letting $h \downarrow 0$, differentiability at x gives

$$f'(x) = \lim_{h \downarrow 0} \frac{f(x+h) - f(x)}{h} \geq 0.$$

If f is decreasing, then correspondingly, for $h > 0$,

$$f(x+h) - f(x) \leq 0,$$

so

$$\frac{f(x+h) - f(x)}{h} \leq 0,$$

and therefore

$$f'(x) \leq 0.$$

□

Remark 425. *The converse implication from "strictly increasing" to "strictly positive derivative" does not hold in general. For example, the function*

$$f(x) = x^3$$

is strictly increasing, but

$$f'(0) = 0.$$

This shows that the condition $f'(x) > 0$ is sufficient, but not necessary, for strict monotonicity.

Theorem 426 (Sufficient criterion for extrema). *Let $f :]a,b[\to \mathbb{R}$ be a differentiable function, and suppose that f is twice differentiable at the point $x \in]a,b[$. Assume that*

$$f'(x) = 0.$$

Then the following statements hold:

- *If $f''(x) > 0$, then f has a strict local minimum at x.*
- *If $f''(x) < 0$, then f has a strict local maximum at x.*

This criterion is only sufficient, but not necessary.

Proof. We consider the case $f''(x) > 0$; the case $f''(x) < 0$ is analogous.

From

$$f''(x) = \lim_{h \to 0} \frac{f'(x+h) - f'(x)}{h} > 0$$

it follows that there exists a $\delta > 0$ such that

$$\frac{f'(x+h) - f'(x)}{h} > 0 \qquad \text{for all } h \neq 0 \text{ with } |h| < \delta.$$

Since $f'(x) = 0$, we obtain from this:

$$f'(x+h) > 0 \quad \text{for } 0 < h < \delta,$$

and

$$f'(x+h) < 0 \quad \text{for } -\delta < h < 0.$$

Now choose δ smaller, if necessary, so that

$$[x-\delta, x+\delta] \subset]a,b[.$$

Then f is continuous on $[x-\delta, x]$ and differentiable on $]x-\delta, x[$. Since $f' < 0$ holds there, f is strictly decreasing on $[x-\delta, x]$. Similarly, f is continuous on $[x, x+\delta]$ and differentiable on $]x, x+\delta[$. Since $f' > 0$ holds there, f is strictly increasing on $[x, x+\delta]$.

Hence, for every $y \in]x-\delta, x[$, we have

$$f(y) > f(x),$$

and likewise for every $y \in]x, x+\delta[$. Thus f has a strict local minimum at x.

The case $f''(x) < 0$ is treated analogously: then f' is positive to the left of x and negative to the right of x, so f has a strict local maximum at x. □

Remark 427. *The condition $f'(x) = 0$ and $f''(x) > 0$ is sufficient, but not necessary, for a local minimum. For example,*

$$f(x) = x^4$$

has a strict local minimum at $x = 0$, but

$$f'(0) = 0 \qquad \textit{and} \qquad f''(0) = 0.$$

Thus an extremum may still occur even when the second derivative vanishes.

Lemma 428. *Let f be differentiable.*

1. *Let $f :]0, a[\to \mathbb{R}$ and*

$$\lim_{x \downarrow 0} f(x) = 0, \qquad \lim_{x \downarrow 0} f'(x) =: c \in \mathbb{R}.$$

Then

$$\lim_{x \downarrow 0} \frac{f(x)}{x} = c.$$

2. *Let $f :]a, \infty[\to \mathbb{R}$ and*

$$\lim_{x \to \infty} f'(x) =: c \in \mathbb{R}.$$

Then

$$\lim_{x \to \infty} \frac{f(x)}{x} = c.$$

Proof. We reduce both statements to the case $c = 0$.

For 1. Set

$$g(x) := f(x) - cx.$$

Then

$$\lim_{x \downarrow 0} g(x) = 0 \qquad \text{and} \qquad \lim_{x \downarrow 0} g'(x) = 0.$$

It is therefore enough to show that from $g(x) \to 0$ and $g'(x) \to 0$ as $x \downarrow 0$ it follows that

$$\frac{g(x)}{x} \to 0.$$

Let $\varepsilon > 0$. Then there exists a $\delta > 0$ such that

$$|g'(t)| < \varepsilon \qquad \text{for all } t \in]0, \delta[.$$

Now fix $x \in]0, \delta[$. Choose $n \in \mathbb{N}$ so large that

$$\frac{x}{2^n} \in]0, \delta[.$$

On the interval $\left[\frac{x}{2^n}, x\right]$, the function g is continuous and differentiable, so the mean value theorem gives a ξ_n such that

$$g(x) - g\left(\frac{x}{2^n}\right) = g'(\xi_n)\left(x - \frac{x}{2^n}\right).$$

Hence

$$\left|g(x) - g\left(\frac{x}{2^n}\right)\right| \leq \varepsilon x.$$

Since $\frac{x}{2^n} \downarrow 0$ as $n \to \infty$ and $g\left(\frac{x}{2^n}\right) \to 0$, passing to the limit gives

$$|g(x)| \leq \varepsilon x.$$

Thus, for all sufficiently small $x > 0$,

$$\left|\frac{g(x)}{x}\right| \leq \varepsilon.$$

Since $\varepsilon > 0$ was arbitrary, it follows that

$$\frac{g(x)}{x} \to 0 \qquad (x \downarrow 0).$$

Consequently,

$$\frac{f(x)}{x} = c + \frac{g(x)}{x} \to c.$$

For 2. Again, set

$$g(x) := f(x) - cx.$$

Then

$$\lim_{x\to\infty} g'(x) = 0.$$

It therefore remains to show that

$$\frac{g(x)}{x} \to 0 \qquad (x \to \infty).$$

Let $\varepsilon > 0$. Then there exists an $R > a$ such that

$$|g'(t)| < \varepsilon \qquad \text{for all } t \geq R.$$

For each $y \geq R$, choose $n \in \mathbb{N}$ such that

$$2^n R \leq y < 2^{n+1} R.$$

On each interval $[2^k R, 2^{k+1} R]$, the mean value theorem yields

$$|g(2^{k+1}R) - g(2^k R)| \leq \varepsilon\, 2^k R.$$

Summing these inequalities gives

$$|g(2^n R)| \leq |g(R)| + \varepsilon R(2^n - 1).$$

Moreover, applying the mean value theorem again on $[2^n R, y]$ gives

$$|g(y) - g(2^n R)| \leq \varepsilon\,(y - 2^n R).$$

Together this implies

$$|g(y)| \leq |g(R)| + \varepsilon R(2^n - 1) + \varepsilon(y - 2^n R) \leq |g(R)| + 2\varepsilon y.$$

Hence

$$\frac{|g(y)|}{y} \leq \frac{|g(R)|}{y} + 2\varepsilon.$$

Letting $y \to \infty$, the first term vanishes. Thus

$$\limsup_{y\to\infty} \frac{|g(y)|}{y} \leq 2\varepsilon.$$

Since $\varepsilon > 0$ was arbitrary, it follows that

$$\frac{g(y)}{y} \to 0.$$

Consequently, here as well,

$$\frac{f(y)}{y} = c + \frac{g(y)}{y} \to c.$$

□

Theorem 429 (Generalized mean value theorem of Cauchy). *Let $f, g : [a, b] \to \mathbb{R}$ be continuous functions that are differentiable on the open interval $]a, b[$. Moreover, assume that*

$$g(b) \neq g(a).$$

Then there exists a point $\phi \in]a, b[$ such that

$$\frac{f'(\phi)}{g'(\phi)} = \frac{f(b) - f(a)}{g(b) - g(a)},$$

provided that $g'(\phi) \neq 0$.

More precisely, without this additional assumption, the equivalent form

$$(f(b) - f(a))g'(\phi) = (g(b) - g(a))f'(\phi).$$

holds.

Proof. The idea is to choose an auxiliary function that takes the same values at the endpoints. Then we can apply Rolle's theorem.

We define

$$h(x) := (f(b) - f(a))\, g(x) - (g(b) - g(a))\, f(x).$$

Since f and g are continuous on $[a, b]$ and differentiable on $]a, b[$, the same is true for h.

We now compute the endpoint values:

$$h(a) = (f(b) - f(a))g(a) - (g(b) - g(a))f(a),$$

and

$$h(b) = (f(b) - f(a))g(b) - (g(b) - g(a))f(b).$$

Subtracting the two expressions, we obtain

$$\begin{aligned} h(b) - h(a) &= (f(b) - f(a))(g(b) - g(a)) \\ &\quad - (g(b) - g(a))(f(b) - f(a)) \\ &= 0. \end{aligned}$$

Thus

$$h(a) = h(b).$$

By Rolle's theorem, there therefore exists a point $\phi \in]a, b[$ such that

$$h'(\phi) = 0.$$

Differentiating h gives

$$h'(x) = (f(b) - f(a))g'(x) - (g(b) - g(a))f'(x).$$

Hence, for ϕ we have

$$(f(b) - f(a))g'(\phi) - (g(b) - g(a))f'(\phi) = 0.$$

Rearranging yields

$$(f(b) - f(a))g'(\phi) = (g(b) - g(a))f'(\phi).$$

This is precisely the desired assertion. □

Remark 430. *Cauchy's theorem is a generalization of the mean value theorem: If we set $g(x) = x$, then*

$$g'(x) = 1, \qquad g(b) - g(a) = b - a,$$

and the ordinary mean value theorem follows immediately from Cauchy's theorem.

Remark 431. *Cauchy's theorem is not only a generalization of the mean value theorem, but also the decisive tool behind l'Hospital's rules. It relates the change of two functions on an interval to the ratio of their derivatives at an interior point. This is precisely what makes it so valuable for limit problems: if the numerator and denominator of a quotient both tend to 0 or both tend* *to ∞, the difficult quotient can often be replaced by a quotient of derivatives.*

In this way, a global question about a limit becomes a local question about the behavior of the derivatives. Thus Cauchy's theorem explains why l'Hospital's rules work in the first place.

13.4 L'Hospital's Rules

L'Hospital's rules are among the most important tools for computing limits. Their basic idea is simple: for quotients whose numerator and denominator both tend to 0 or both grow without bound, the quotient of the derivatives is often easier to analyze than the original quotient. The generalized mean value theorem makes this idea precise.

Theorem 432 (L'Hospital's rules). *Let f, g be differentiable on a suitable interval.*

(i) ***The case*** $0/0$. *Let $a \in \mathbb{R}$ and let $f, g : [a, b[\to \mathbb{R}$ be continuous at a and differentiable on $]a, b[$. Suppose that*

$$f(a) = g(a) = 0,$$
$$g(x) \neq 0 \quad \text{for } x > a \text{ sufficiently close to } a,$$

and suppose moreover that

$$\lim_{x \downarrow a} \frac{f'(x)}{g'(x)} = L \in \mathbb{R}.$$

Then

$$\lim_{x \downarrow a} \frac{f(x)}{g(x)} = L.$$

(ii) ***The case*** ∞/∞. *Let* $f, g :]A, \infty[\to \mathbb{R}$ *be differentiable and suppose that*

$$\lim_{x \to \infty} f(x) = \infty, \qquad \lim_{x \to \infty} g(x) = \infty.$$

Moreover, assume that $g(x) \neq 0$ *for all sufficiently large* x *and that*

$$\lim_{x \to \infty} \frac{f'(x)}{g'(x)} = L \in \mathbb{R}.$$

Then

$$\lim_{x \to \infty} \frac{f(x)}{g(x)} = L.$$

Proof. We use the generalized mean value theorem, also known as Cauchy's mean value theorem.

Proof of (i). Since $f(a) = g(a) = 0$ and f, g are continuous at a, for $x \in]a, b[$ sufficiently close to a we consider the functions on the interval $[a, x]$. By Cauchy's mean value theorem, there exists a point $\xi_x \in]a, x[$ such that

$$\frac{f(x) - f(a)}{g(x) - g(a)} = \frac{f'(\xi_x)}{g'(\xi_x)}.$$

Since $f(a) = g(a) = 0$, this simplifies to

$$\frac{f(x)}{g(x)} = \frac{f'(\xi_x)}{g'(\xi_x)}.$$

Now $\xi_x \in]a, x[$, hence $\xi_x \downarrow a$ whenever $x \downarrow a$. Since, by assumption,

$$\frac{f'(t)}{g'(t)} \to L \qquad (t \downarrow a),$$

it follows immediately that

$$\frac{f(x)}{g(x)} \to L \qquad (x \downarrow a).$$

Proof of (ii). Set

$$h(x) := f(x) - Lg(x).$$

Then

$$\frac{h'(x)}{g'(x)} = \frac{f'(x) - Lg'(x)}{g'(x)} = \frac{f'(x)}{g'(x)} - L \to 0 \qquad (x \to \infty).$$

It therefore suffices to show that

$$\frac{h(x)}{g(x)} \to 0.$$

Let $\varepsilon > 0$. Then there exists $R > A$ such that

$$\left|\frac{h'(x)}{g'(x)}\right| < \varepsilon \qquad \text{for all } x \geq R.$$

For $x > R$, we apply Cauchy's mean value theorem to the interval $[R, x]$. Then there exists a point $\eta_x \in]R, x[$ such that

$$\frac{h(x) - h(R)}{g(x) - g(R)} = \frac{h'(\eta_x)}{g'(\eta_x)}.$$

Since $\eta_x > R$, it follows that

$$\left|\frac{h(x) - h(R)}{g(x) - g(R)}\right| < \varepsilon.$$

We now write

$$\frac{h(x)}{g(x)} = \frac{h(x) - h(R)}{g(x) - g(R)} \cdot \frac{g(x) - g(R)}{g(x)} + \frac{h(R)}{g(x)}.$$

Since $g(x) \to \infty$, we have

$$\frac{g(x) - g(R)}{g(x)} \to 1 \qquad \text{and} \qquad \frac{h(R)}{g(x)} \to 0.$$

Hence

$$\limsup_{x \to \infty} \left|\frac{h(x)}{g(x)}\right| \le \varepsilon.$$

Since $\varepsilon > 0$ was arbitrary, we obtain

$$\frac{h(x)}{g(x)} \to 0.$$

Thus

$$\frac{f(x)}{g(x)} = \frac{Lg(x) + h(x)}{g(x)} = L + \frac{h(x)}{g(x)} \to L.$$

This proves the claim. □

Remark 433. *The case $g(x) \to -\infty$ is analogous; by multiplying numerator and denominator by -1, it can be reduced to the case $g(x) \to \infty$.*

13.5 Examples of the Rules of de l'Hospital

The rules of de l'Hospital are especially useful for limits that initially appear to be indeterminate. Typical examples are expressions of the form $0/0$ or ∞/∞. In such cases, one replaces the original quotient by the quotient of the derivatives. The following examples illustrate the basic idea in simple but important situations.

Example 434 (The limit of $\sin x/x$). *We have*

$$\lim_{x\to 0} \frac{\sin x}{x} = 1.$$

Proof. As $x \to 0$, numerator and denominator form an indeterminate expression of type $0/0$. Since

$$(\sin x)' = \cos x \qquad \text{and} \qquad (x)' = 1,$$

de l'Hospital's rule gives

$$\lim_{x\to 0} \frac{\sin x}{x} = \lim_{x\to 0} \frac{\cos x}{1} = \cos 0 = 1.$$

□

Remark 435. *This limit is one of the most important standard limits in analysis. In particular, it shows that near* 0*, the sine function grows very much like the identity function* $x \mapsto x$.

Example 436 (The limit of $\ln(1+x)/x$). *We have*

$$\lim_{x\to 0}\frac{\ln(1+x)}{x}=1.$$

Proof. Here too, we have a case of the form $0/0$, since

$$\ln(1+0)=0.$$

Moreover,

$$(\ln(1+x))'=\frac{1}{1+x}\qquad\text{and}\qquad (x)'=1.$$

Hence, by l'Hospital's rule,

$$\lim_{x\to 0}\frac{\ln(1+x)}{x}=\lim_{x\to 0}\frac{1}{1+x}=1.$$

□

Remark 437. *This limit shows that the logarithm is almost linear near* 1*. More precisely, for small* x*, the function* $\ln(1+x)$ *behaves like* x *itself.*

Example 438 (Comparison of the exponential function and powers). *Let* $n\in\mathbb{N}$*. Then*

$$\lim_{x\to\infty}\frac{e^x}{x^n}=\infty.$$

Proof. First, we have the indeterminate form ∞/∞. We apply l'Hospital's rule repeatedly:

$$\begin{aligned}\lim_{x\to\infty}\frac{e^x}{x^n} &= \lim_{x\to\infty}\frac{e^x}{nx^{n-1}}\\ &= \lim_{x\to\infty}\frac{e^x}{n(n-1)x^{n-2}}\\ &= \cdots\\ &= \lim_{x\to\infty}\frac{e^x}{n!}.\end{aligned}$$

Since $e^x \to \infty$ as $x \to \infty$, it follows that

$$\lim_{x\to\infty}\frac{e^x}{x^n} = \infty.$$

□

Remark 439. *This example makes clear that the exponential function dominates every power. This is a fundamental growth rule that will reappear later in many contexts.*

Example 440 (Another typical application)**.** *We have*

$$\lim_{x\to 0}\frac{e^x-1}{x} = 1.$$

Proof. Again we have an expression of the indeterminate form $0/0$. Since

$$(e^x-1)' = e^x \qquad \text{and} \qquad (x)' = 1$$

we obtain

$$\lim_{x\to 0}\frac{e^x-1}{x} = \lim_{x\to 0} e^x = 1.$$

□

13.6 Exercises and Solutions

Exercise 441. *Investigate the function*

$$f(x) = x^3 - 3x$$

for local extrema and determine them.

Solution 442. *We have*

$$f'(x) = 3x^2 - 3 = 3(x-1)(x+1).$$

Thus the critical points are $x = -1$ *and* $x = 1$.

For classification, we compute the second derivative:

$$f''(x) = 6x.$$

Hence

$$f''(-1) = -6 < 0,$$

so f *has a strict local maximum at* $x = -1$. *Furthermore,*

$$f''(1) = 6 > 0,$$

so f *has a strict local minimum at* $x = 1$.

The corresponding function values are

$$f(-1) = 2, \qquad f(1) = -2.$$

Exercise 443. *Use Rolle's theorem to show that a polynomial of degree* n *can have at most* n *distinct real zeros.*

Solution 444. *We prove the statement by induction on n.*

For $n = 0$, the assertion is clear, since a constant nonzero polynomial has no zeros at all.

Now assume that the statement has already been proved for polynomials of degree $n-1$, and let p be a polynomial of degree n with $n+1$ distinct real zeros

$$x_0 < x_1 < \cdots < x_n.$$

Then, on each interval $[x_{k-1}, x_k]$, we have

$$p(x_{k-1}) = p(x_k) = 0.$$

By Rolle's theorem, for each $k = 1, \ldots, n$ there is therefore a point

$$\phi_k \in]x_{k-1}, x_k[$$

such that

$$p'(\phi_k) = 0.$$

Thus p' has at least n distinct real zeros. Since p' is a polynomial of degree $n-1$, this contradicts the induction hypothesis. Hence p can have at most n distinct real zeros.

Exercise 445. *Show that, for $0 < x < \frac{\pi}{2}$, the following estimate holds:*

$$\sin x < x < \tan x.$$

Solution 446. *For the first inequality, consider the function* sin *on* $[0, x]$. *By the mean value theorem, there exists a point* $\xi \in]0, x[$ *such that*

$$\sin x - \sin 0 = \cos(\xi)\,(x - 0).$$

Hence

$$\sin x = x \cos(\xi).$$

Since $\xi \in]0, \frac{\pi}{2}[$, *we have* $\cos(\xi) < 1$, *and therefore*

$$\sin x < x.$$

For the second inequality, consider the function tan *on* $[0, x]$. *Again, the mean value theorem gives a point* $\eta \in]0, x[$ *such that*

$$\tan x - \tan 0 = \sec^2(\eta)\, x.$$

Hence

$$\tan x = x \sec^2(\eta).$$

Since $\sec^2(\eta) > 1$, *it follows that*

$$\tan x > x.$$

Together, this yields

$$\sin x < x < \tan x.$$

Exercise 447. *Show that, for all* $x > -1$, *the inequality*

$$\ln(1 + x) \leq x,$$

holds, and determine the case of equality.

Solution 448. *Consider*

$$g(x) = x - \ln(1+x), \qquad x > -1.$$

Then

$$g'(x) = 1 - \frac{1}{1+x} = \frac{x}{1+x}.$$

For $-1 < x < 0$ *we have* $g'(x) < 0$*, and for* $x > 0$ *we have* $g'(x) > 0$*. Hence* g *is decreasing on* $(-1, 0]$ *and increasing on* $[0, \infty)$*. Therefore* g *has its global minimum at* $x = 0$*.*

Moreover,

$$g(0) = 0.$$

It follows that

$$g(x) \geq 0,$$

that is,

$$\ln(1+x) \leq x$$

for all $x > -1$*.*

Equality holds if and only if $x = 0$*.*

Exercise 449. *Give an example of a strictly increasing function whose derivative vanishes at one point.*

Solution 450. *A standard example is*

$$f(x) = x^3.$$

This function is strictly increasing on all of $\mathbb{R}$*, since*

$$x < y \implies x^3 < y^3.$$

On the other hand,

$$f'(x) = 3x^2,$$

and in particular

$$f'(0) = 0.$$

This shows that strict monotonicity does not require the condition $f'(x) > 0$ everywhere.

Exercise 451. *Investigate the function*

$$f(x) = x^4 - 2x^2$$

for local extrema.

Solution 452. *We have*

$$f'(x) = 4x^3 - 4x = 4x(x^2 - 1).$$

Thus the critical points are

$$x = -1,\ 0,\ 1.$$

Furthermore,

$$f''(x) = 12x^2 - 4.$$

It follows that

$$f''(0) = -4 < 0,$$

so f has a strict local maximum at $x = 0$.

Moreover,

$$f''(\pm 1) = 12 - 4 = 8 > 0,$$

so f has a strict local minimum at both $x = -1$ and $x = 1$.

The function values are

$$f(0) = 0, \qquad f(\pm 1) = -1.$$

Exercise 453. *Compute the limit*

$$\lim_{x \to 0} \frac{e^x - 1 - x}{x^2}.$$

Solution 454. *This is of the indeterminate form 0/0. We apply l'Hospital's rule twice:*

$$\lim_{x \to 0} \frac{e^x - 1 - x}{x^2} = \lim_{x \to 0} \frac{e^x - 1}{2x} = \lim_{x \to 0} \frac{e^x}{2} = \frac{1}{2}.$$

Exercise 455. *Compute the limit*

$$\lim_{x \to 0} \frac{1 - \cos x}{x^2}.$$

Solution 456. *Here, too, we have the indeterminate form 0/0. By l'Hospital's rule, we obtain*

$$\lim_{x \to 0} \frac{1 - \cos x}{x^2} = \lim_{x \to 0} \frac{\sin x}{2x}.$$

The last limit is again of the form 0/0, so we apply l'Hospital's rule once more:

$$\lim_{x \to 0} \frac{\sin x}{2x} = \lim_{x \to 0} \frac{\cos x}{2} = \frac{1}{2}.$$

Exercise 457. *Compute the limit*

$$\lim_{x\to\infty} \frac{\ln x}{x}.$$

Solution 458. *This is an indeterminate form of type* ∞/∞*. By l'Hospital's rule, we obtain*

$$\lim_{x\to\infty} \frac{\ln x}{x} = \lim_{x\to\infty} \frac{1/x}{1} = \lim_{x\to\infty} \frac{1}{x} = 0.$$

Exercise 459. *Let* $f : [a,b] \to \mathbb{R}$ *be continuous and differentiable on* $]a,b[$*. Suppose that*

$$|f'(x)| \leq M \qquad \text{for all } x \in]a,b[.$$

Show that, for all $x, y \in [a,b]$*, one then has*

$$|f(y) - f(x)| \leq M|y - x|.$$

Solution 460. *Let* $x, y \in [a,b]$ *with* $x < y$*. By the mean value theorem, there exists a point* $\phi \in]x,y[$ *such that*

$$f'(\phi) = \frac{f(y) - f(x)}{y - x}.$$

It follows that

$$\left|\frac{f(y) - f(x)}{y - x}\right| = |f'(\phi) \ \leq M.$$

Multiplication by $y - x > 0$ *gives*

$$|f(y) - f(x)| \leq M(y - x).$$

For $x = y$, the assertion is trivial. Since for $x > y$ the roles of x and y can simply be interchanged, we obtain in general

$$|f(y) - f(x)| \leq M|y - x|.$$

Exercise 461. *Let $c \in \mathbb{R}$ be a constant and let $f : \mathbb{R} \to \mathbb{R}$ be a differentiable function satisfying*

$$f'(x) = c\,f(x) \qquad \text{for all } x \in \mathbb{R}.$$

Show that then

$$f(x) = f(0)e^{cx} \qquad \text{for all } x \in \mathbb{R}$$

holds.

Solution 462. *The idea is to rewrite the function in such a way that its derivative vanishes. To this end, we consider*

$$g(x) := e^{-cx} f(x).$$

Since f and the exponential function are differentiable, g is differentiable as well. By the product rule, we obtain

$$g'(x) = (-c)e^{-cx} f(x) + e^{-cx} f'(x).$$

Now we use the assumption $f'(x) = c\,f(x)$. This yields

$$g'(x) = -ce^{-cx} f(x) + e^{-cx} \cdot cf(x) = 0$$

for all $x \in \mathbb{R}$.

Thus g is constant. In particular,

$$g(x) = g(0) \qquad \text{for all } x \in \mathbb{R}.$$

From the definition of g, it follows that

$$e^{-cx}f(x) = f(0),$$

since

$$g(0) = e^0 f(0) = f(0).$$

Multiplication by e^{cx} *finally gives*

$$f(x) = f(0)e^{cx}$$

for all $x \in \mathbb{R}$.

This proves the assertion.

Remark 463. *The preceding exercise shows that the differential equation*

$$f'(x) = c\,f(x)$$

already determines a function uniquely as soon as its value at one point is known.

In particular, for $c = 1$ *and the initial condition*

$$f(0) = 1$$

one obtains the solution

$$f(x) = e^x.$$

Thus the exponential function can be characterized as the uniquely determined differentiable function

$$f : \mathbb{R} \to \mathbb{R}$$

that satisfies the conditions

$$f'(x) = f(x) \qquad \textit{and} \qquad f(0) = 1$$

.

This property makes the exponential function one of the most important functions in analysis and explains its frequent occurrence in the natural sciences, engineering, and probability theory.

Chapter 14

Integration

After introducing the derivative in the preceding chapters as a local measure of the rate of change of a function, we now turn to a second fundamental idea of analysis: integration. While differentiation asks for the slope of a graph, integration is concerned with a notion of area. These two points of view are inseparably connected and together form the core of analysis.

The intuitive idea is simple. One considers the region under the graph of a function on an interval $[a, b]$ and wants to determine its area. For rectangles this can be done immediately: the area is the product of width and height. For more general functions, however, a direct calculation is usually no longer possible. We therefore first approximate the desired area by suitable rectangular sums.

This is precisely where *step functions* come into

play. They form a particularly simple class of functions composed of finitely many rectangles. For them, an integral can be defined without taking limits. This simple situation is not only computationally manageable, but also didactically valuable: it makes clear what the integral is meant to accomplish before we move on to more general functions and ultimately to limiting processes.

14.1 Step Functions

We begin with a class of functions that consist of piecewise constant parts. Such functions resemble a staircase: on each subinterval their value remains constant, and only at finitely many points do they jump to a new value.

Definition 464 (Step function). *A function* $f : [a, b] \to \mathbb{R}$ *is called a* step function *if there exists a partition*

$$a = x_0 < x_1 < \cdots < x_n = b$$

of the interval $[a, b]$ *such that* f *is constant on each subinterval*

$$[x_{k-1}, x_k) \qquad (k = 1, \ldots, n)$$

Thus, for $k = 1, \ldots, n$*, there are numbers* $c_k \in \mathbb{R}$ *such that*

$$f(x) = c_k \quad \text{for all } x \in [x_{k-1}, x_k).$$

At the right endpoint b*, the function value may, if necessary, be specified separately.*

Step functions are the simplest functions for which an area calculation can be carried out in a meaningful way. Indeed, on each subinterval the graph is horizontal, and therefore its contribution to the area is precisely the area of a rectangle.

Proposition 465. *The set $T[a, b]$ of all step functions on $[a, b]$ is a vector subspace of the vector space $\mathbb{R}^{[a,b]}$ of all real-valued functions on $[a, b]$.*

Proof. The zero function is a step function, for instance with respect to any partition of $[a, b]$.

Now let $f, g \in T[a, b]$. Then there exist partitions

$$a = x_0 < \cdots < x_n = b \qquad \text{and} \qquad a = y_0 < \cdots < y_m = b$$

as well as numbers $c_1, \ldots, c_n$ and $d_1, \ldots, d_m$ such that

$$\begin{aligned} f(x) &= c_j \quad \text{for } x \in [x_{j-1}, x_j), \\ g(x) &= d_\ell \quad \text{for } x \in [y_{\ell-1}, y_\ell). \end{aligned}$$

Consider the common refinement

$$a = z_0 < z_1 < \cdots < z_r = b$$

of all points $x_0, \ldots, x_n, y_0, \ldots, y_m$. Then each interval $[z_{k-1}, z_k)$ lies in exactly one of the intervals $[x_{j-1}, x_j)$ and in exactly one of the intervals $[y_{\ell-1}, y_\ell)$. Consequently, on each $[z_{k-1}, z_k)$, both f and g are constant, and hence so is

$$(f + g)(x) = c_j + d_\ell \quad \text{for } x \in [z_{k-1}, z_k).$$

Thus $f + g$ is again a step function.

If $\lambda \in \mathbb{R}$, then on the same interval we have

$$(\lambda f)(x) = \lambda c_j \quad \text{for } x \in [z_{k-1}, z_k),$$

so λf is also a step function.

Hence $T[a, b]$ is closed under addition and scalar multiplication and contains the zero function. Therefore $T[a, b]$ is a vector subspace of $\mathbb{R}^{[a,b]}$. □

Definition 466 (Integral of a step function). *Let $f : [a, b] \to \mathbb{R}$ be a step function, and let*

$$a = x_0 < x_1 < \cdots < x_n = b$$

be a partition on whose subintervals f is constant, with

$$f(x) = c_k \qquad \text{for } x \in [x_{k-1}, x_k).$$

Then we define the integral *of f over $[a, b]$ by*

$$\int_a^b f(x)\,dx := \sum_{k=1}^{n} c_k\,(x_k - x_{k-1}).$$

This definition corresponds exactly to the idea of computing the area under the graph as a sum of rectangular areas. The term $x_k - x_{k-1}$ is the width of the k-th rectangle, and the factor c_k is its height.

Remark 467. *If a step function is represented with respect to different partitions, the same area gives the same integral sum. Thus the integral is well defined and does not depend on the chosen partition, but only on the function itself.*

Proposition 468. *The integral of a step function is independent of the chosen partition.*

Proof. Let $f : [a, b] \to \mathbb{R}$ be a step function. Suppose that f is given with respect to the partition

$$a = x_0 < x_1 < \cdots < x_n = b$$

by

$$f(x) = c_j \qquad \text{for } x \in [x_{j-1}, x_j)$$

and with respect to the partition

$$a = y_0 < y_1 < \cdots < y_m = b$$

by

$$f(x) = d_\ell \qquad \text{for } x \in [y_{\ell-1}, y_\ell).$$

We consider the common refinement

$$a = z_0 < z_1 < \cdots < z_r = b$$

of all points $x_0, \ldots, x_n, y_0, \ldots, y_m$.

Now let $I_j = [x_{j-1}, x_j)$. Then the points of the refinement divide I_j into finitely many subintervals

$$[z_{k-1}, z_k) \subset I_j$$

and on each of these subintervals one again has $f = c_j$. Therefore,

$$\begin{aligned} \sum_{[z_{k-1}, z_k) \subset I_j} c_j \,(z_k - z_{k-1}) &= c_j \sum_{[z_{k-1}, z_k) \subset I_j} (z_k - z_{k-1}) \\ &= c_j (x_j - x_{j-1}). \end{aligned}$$

Summing over all j, it follows that

$$\sum_{k=1}^{r} \tilde{c}_k \,(z_k - z_{k-1}) = \sum_{j=1}^{n} c_j \,(x_j - x_{j-1}),$$

where $\tilde{c}_k$ is the constant value of f on $[z_{k-1}, z_k)$.

Analogously, one obtains

$$\sum_{k=1}^{r} \tilde{c}_k \,(z_k - z_{k-1}) = \sum_{\ell=1}^{m} d_\ell \,(y_\ell - y_{\ell-1}).$$

Hence

$$\sum_{j=1}^{n} c_j \,(x_j - x_{j-1}) = \sum_{\ell=1}^{m} d_\ell \,(y_\ell - y_{\ell-1}).$$

Thus the value of the integral is independent of the chosen partition. □

Example 469. *Let* $f : [0,3] \to \mathbb{R}$ *be given by*

$$f(x) = \begin{cases} 2, & x \in [0,1), \\ 1, & x \in [1,2), \\ 3, & x \in [2,3]. \end{cases}$$

Then f *is a step function, and*

$$\int_0^3 f(x)\,dx = 2\cdot(1-0)+1\cdot(2-1)+3\cdot(3-2) = 2+1+3 = 6.$$

Step functions therefore provide the first access to the integral. In the following sections, we shall use them systematically in order to extend the concept of the integral step by step to broader classes of functions. The geometric intuition developed in this section will continue to play a central role.

Before extending the notion of the integral to more general classes of functions, it is useful to establish the basic computational rules already in the simple case of step functions.

Intuitively, this is clear: the integral should behave like a sum of areas. If two areas are considered together, their contents add; if a function is scaled, then the corresponding area is scaled as well; and a larger function should also produce a larger area under its graph.

Precisely these properties can be proved completely within the framework of step functions, without using any limiting processes. They thus form the algebraic foundation of the later concept of the integral.

Theorem 470. *Let $\phi, \psi \in T[a,b]$ and $\lambda \in \mathbb{R}$. Then:*

1. $\displaystyle\int_a^b (\phi + \psi)(x)\,dx = \int_a^b \phi(x)\,dx + \int_a^b \psi(x)\,dx.$

2. $\displaystyle\int_a^b (\lambda\phi)(x)\,dx = \lambda \int_a^b \phi(x)\,dx.$

3. *If $\phi \le \psi$, then* $\displaystyle\int_a^b \phi(x)\,dx \le \int_a^b \psi(x)\,dx.$

Here

$$\phi \le \psi \quad \Longleftrightarrow \quad \phi(x) \le \psi(x) \text{ for all } x \in [a,b].$$

Proof. For the two functions, we first choose a common partition

$$a = x_0 < x_1 < \cdots < x_n = b,$$

such that both ϕ and ψ are constant on each subinterval $[x_{k-1}, x_k)$. We then write

$$\phi(x) = \alpha_k, \qquad \psi(x) = \beta_k \quad \text{for } x \in [x_{k-1}, x_k).$$

Then $\phi + \psi$ is also constant on each subinterval, with value $\alpha_k + \beta_k$, and likewise $\lambda\phi$ has value $\lambda\alpha_k$.

Proof of 1. By the definition of the integral,

$$\int_a^b (\phi + \psi)(x)\,dx = \sum_{k=1}^{n} (\alpha_k + \beta_k)(x_k - x_{k-1}).$$

Using the rules for finite sums, we obtain

$$\begin{aligned}
&\sum_{k=1}^{n} (\alpha_k + \beta_k)(x_k - x_{k-1}) \\
&\quad = \sum_{k=1}^{n} \alpha_k (x_k - x_{k-1}) \\
&\qquad + \sum_{k=1}^{n} \beta_k (x_k - x_{k-1}) \\
&\quad = \int_a^b \phi(x)\,dx + \int_a^b \psi(x)\,dx.
\end{aligned}$$

This proves 1.

For 2. Similarly, we have

$$\int_a^b (\lambda\phi)(x)\,dx = \sum_{k=1}^{n} (\lambda\alpha_k)(x_k - x_{k-1}).$$

Since the scalar λ can be factored out of a sum, it follows that

$$\sum_{k=1}^{n} (\lambda\alpha_k)(x_k - x_{k-1}) = \lambda \sum_{k=1}^{n} \alpha_k(x_k - x_{k-1}) = \lambda \int_a^b \phi(x)\,dx.$$

Thus 2. also holds.

For 3. From $\phi \leq \psi$, it follows for all k that

$$\alpha_k \leq \beta_k.$$

Hence also

$$\beta_k - \alpha_k \geq 0.$$

Now consider the difference function $\psi - \phi$. It is again a step function, and on $[x_{k-1}, x_k)$ we have

$$(\psi - \phi)(x) = \beta_k - \alpha_k \geq 0.$$

Its integral is therefore

$$\int_a^b (\psi - \phi)(x)\,dx = \sum_{k=1}^{n} (\beta_k - \alpha_k)(x_k - x_{k-1}) \geq 0,$$

because all summands are nonnegative.

Using Part 1 and Part 2, we now obtain

$$\int_a^b \psi(x)\,dx - \int_a^b \phi(x)\,dx = \int_a^b (\psi - \phi)(x)\,dx \geq 0.$$

Thus
$$\int_a^b \phi(x)\,dx \leq \int_a^b \psi(x)\,dx.$$
This proves the assertion. □

14.2 Upper and Lower Integral

Step functions provide a first class of functions for which the notion of the integral can be defined directly and without limiting processes. The next step is to approximate general bounded functions from below and from above by step functions.

The basic idea is simple: A step function φ with $\varphi \leq f$ provides a lower estimate for the area associated with f, while a step function ψ with $\psi \geq f$ provides an upper estimate. The closer these two bounds move toward each other, the more accurately they capture the desired area under the graph of f.

Definition of Upper and Lower Integral

Definition 471 (Bounded function). *A function $f : [a, b] \to \mathbb{R}$ is called* bounded *if there exist numbers $m, M \in \mathbb{R}$ such that*

$$m \leq f(x) \leq M \qquad \textit{for all } x \in [a, b].$$

In what follows, we consider only bounded functions. For such functions, step functions can always be used as lower and upper bounds.

Definition 472 (Lower and upper integral bounds). *Let $f : [a,b] \to \mathbb{R}$ be a bounded function. We consider the sets*

$$L(f) := \left\{ \int_a^b \varphi(x)\,dx \;:\; \varphi \in T[a,b],\ \varphi \leq f \right\}$$

and

$$U(f) := \left\{ \int_a^b \psi(x)\,dx \;:\; \psi \in T[a,b],\ \psi \geq f \right\}.$$

Then we call

$$\underline{\int_a^b} f(x)\,dx := \sup L(f)$$

the lower integral *of f over $[a,b]$, and*

$$\overline{\int_a^b} f(x)\,dx := \inf U(f)$$

the upper integral *of f over $[a,b]$.*

This definition is very natural in meaning: The lower integral is the greatest area value that can be attained by step functions from below. The upper integral is the smallest area value obtained from step functions from above.

Remark 473. *For step functions themselves, this definition agrees with the integral already introduced. This shows precisely that the integral of a step function is the starting point of the general theory of integration.*

Existence of Upper and Lower Bounds

Proposition 474. *If $f : [a, b] \to \mathbb{R}$ is bounded, then $L(f)$ and $U(f)$ are nonempty. Moreover, $L(f)$ is bounded above and $U(f)$ is bounded below.*

Proof. Since f is bounded, there exist numbers $m, M \in \mathbb{R}$ such that

$$m \leq f(x) \leq M \qquad \text{for all } x \in [a, b].$$

The constant functions $\varphi(x) = m$ and $\psi(x) = M$ are step functions. Since

$$\varphi \leq f \leq \psi$$

we have $\int_a^b \varphi(x)\,dx \in L(f)$ and $\int_a^b \psi(x)\,dx \in U(f)$. Hence both sets are nonempty.

Now let $\eta \in T[a, b]$ with $\eta \leq f$. Then also $\eta \leq M$, and therefore, by the monotonicity of the integral for step functions,

$$\int_a^b \eta(x)\,dx \leq \int_a^b M\,dx = M(b-a).$$

Thus $L(f)$ is bounded above.

Analogously, for every step function $\theta \geq f$, it follows that $\theta \geq m$, hence

$$\int_a^b \theta(x)\,dx \geq \int_a^b m\,dx = m(b-a).$$

Thus $U(f)$ is bounded below. □

The lower integral never exceeds the upper integral

Proposition 475. *For every bounded function* $f : [a,b] \to \mathbb{R}$*, one has*

$$\underline{\int_a^b} f(x)\,dx \le \overline{\int_a^b} f(x)\,dx.$$

Proof. Let $\varphi \in T[a,b]$ with $\varphi \le f$, and let $\psi \in T[a,b]$ with $\psi \ge f$. Then

$$\varphi \le \psi.$$

By the monotonicity of the integral for step functions, it follows that

$$\int_a^b \varphi(x)\,dx \le \int_a^b \psi(x)\,dx.$$

Since this holds for every such pair (φ,ψ), every number in $L(f)$ is at most as large as every number in $U(f)$. Hence, in particular,

$$\sup L(f) \le \inf U(f),$$

and therefore

$$\underline{\int_a^b} f(x)\,dx \le \overline{\int_a^b} f(x)\,dx.$$

□

14.2.1 Integrability

Definition 476 (Riemann integrability). *A bounded function* $f : [a, b] \to \mathbb{R}$ *is called* integrable *if*

$$\underline{\int_a^b} f(x)\,dx = \overline{\int_a^b} f(x)\,dx.$$

In this case, we simply write the common value as

$$\int_a^b f(x)\,dx.$$

Thus, for more general functions, the notion of the integral is no longer defined directly as the area of individual rectangles, but rather as a limiting or bounding value determined by step functions.

Relationship with the integral of step functions

Theorem 477. *Every step function is integrable. For a step function* $f \in T[a, b]$*, the lower integral, the upper integral, and the previously defined integral coincide:*

$$\underline{\int_a^b} f(x)\,dx = \overline{\int_a^b} f(x)\,dx = \int_a^b f(x)\,dx.$$

Proof. Let $f \in T[a, b]$. Then, of course,

$$f \leq f \qquad \text{and} \qquad f \geq f.$$

Thus f is at the same time an admissible function for the lower integral and for the upper integral.

From the definition of $L(f)$, it follows that

$$\int_a^b f(x)\,dx \in L(f),$$

hence

$$\underline{\int_a^b} f(x)\,dx = \sup L(f) \geq \int_a^b f(x)\,dx.$$

Similarly,

$$\int_a^b f(x)\,dx \in U(f),$$

hence

$$\overline{\int_a^b} f(x)\,dx = \inf U(f) \leq \int_a^b f(x)\,dx.$$

On the other hand, by the preceding proposition we always have

$$\underline{\int_a^b} f(x)\,dx \leq \overline{\int_a^b} f(x)\,dx.$$

Combining these inequalities gives

$$\underline{\int_a^b} f(x)\,dx \leq \overline{\int_a^b} f(x)\,dx \leq \int_a^b f(x)\,dx \leq \underline{\int_a^b} f(x)\,dx.$$

Therefore, all three numbers must be equal. □

Examples

Example 478 (Constant function). *Let $f(x) = c$ for all $x \in [a, b]$. Then f is already a step function. Hence*

$$\underline{\int_a^b} f(x)\,dx = \overline{\int_a^b} f(x)\,dx = \int_a^b f(x)\,dx = c(b-a).$$

Example 479 (The function $f(x) = x$ on [0,1]). *We consider the partition*

$$0 = \frac{0}{n} < \frac{1}{n} < \cdots < \frac{n}{n} = 1.$$

Define two step functions φ_n, ψ_n by

$$\varphi_n(x) = \frac{k}{n} \quad \text{for } x \in \left[\frac{k}{n}, \frac{k+1}{n}\right),$$
$$\psi_n(x) = \frac{k+1}{n} \quad \text{for } x \in \left[\frac{k}{n}, \frac{k+1}{n}\right).$$

for $k = 0, \ldots, n-1$. Then, on each subinterval,

$$\varphi_n(x) \le x \le \psi_n(x).$$

Thus

$$\varphi_n \le f \le \psi_n.$$

For the integrals, we obtain

$$\int_0^1 \varphi_n(x)\,dx = \sum_{k=0}^{n-1} \frac{k}{n} \cdot \frac{1}{n} = \frac{1}{n^2} \sum_{k=0}^{n-1} k = \frac{n-1}{2n},$$

and

$$\int_0^1 \psi_n(x)\,dx = \sum_{k=0}^{n-1} \frac{k+1}{n} \cdot \frac{1}{n} = \frac{1}{n^2} \sum_{k=1}^{n} k = \frac{n+1}{2n}.$$

It follows that

$$\underline{\int_0^1} x\,dx \geq \frac{n-1}{2n} \qquad \text{and} \qquad \overline{\int_0^1} x\,dx \leq \frac{n+1}{2n}$$

for all $n \in \mathbb{N}$*. Letting* $n \to \infty$*, we obtain*

$$\underline{\int_0^1} x\,dx \geq \frac{1}{2} \qquad \text{and} \qquad \overline{\int_0^1} x\,dx \leq \frac{1}{2}.$$

Together with

$$\underline{\int_0^1} x\,dx \leq \overline{\int_0^1} x\,dx$$

this implies

$$\underline{\int_0^1} x\,dx = \overline{\int_0^1} x\,dx = \frac{1}{2}.$$

Thus the function $f(x) = x$ *is integrable, and its integral is* $\frac{1}{2}$.

Example 480 (The Dirichlet function). *We consider the function*

$$f(x) = \begin{cases} 1, & x \in \mathbb{Q}, \\ 0, & x \notin \mathbb{Q}, \end{cases} \qquad x \in [0,1].$$

This function is bounded, but it is not integrable.

Indeed, every step function φ with $\varphi \leq f$ must be constant on each subinterval. If its value on some subinterval were greater than 0, then, since irrational numbers occur in every interval, there would also be points there with $f = 0$, and the condition $\varphi \leq f$ would be violated. Hence necessarily $\varphi \leq 0$, and therefore

$$\underline{\int_0^1} f(x)\,dx = 0.$$

Similarly, every step function ψ with $\psi \geq f$ must have value at least 1 on each subinterval, since rational numbers occur in every interval. Hence

$$\overline{\int_0^1} f(x)\,dx = 1.$$

Thus the lower integral and the upper integral are different, and so f is not integrable.

Summary 481. *Step functions form the starting point of integration. For them, the integral is defined directly as a sum of rectangular areas. For general bounded functions, we extend this concept by approximation from below and from above:*

- *The* lower integral *is the largest area attained by step functions from below.*
- *The* upper integral *is the smallest area attained by step functions from above.*

- *If the two values agree, then the function is integrable.*
- *For step functions themselves, this new definition coincides with the already known integral.*

In this way, the concept of the integral is constructed not only geometrically, but also in a logically precise manner: first, simple functions are integrated exactly; then general functions are described by means of these simple building blocks.

14.2.2 Riemann Integrability

Theorem 482 (Criterion for Riemann integrability). *A function $f : [a, b] \to \mathbb{R}$ is Riemann integrable if and only if, for every $\epsilon > 0$, there exist step functions $\varphi, \psi \in T[a, b]$ such that*

$$\varphi \leq f \leq \psi$$

and

$$\int_a^b \psi(x)\,dx - \int_a^b \varphi(x)\,dx < \epsilon.$$

Proof. We set

$$L := \underline{\int_a^b} f(x)\,dx \qquad \text{and} \qquad U := \overline{\int_a^b} f(x)\,dx.$$

Then always $L \leq U$.

$\Rightarrow$: Suppose that f is Riemann integrable. Then $L = U$. Let $\epsilon > 0$ be given. By the definition of the supremum, there exists a step function $\varphi \in T[a,b]$ with $\varphi \leq f$ and

$$L - \int_a^b \varphi(x)\,dx < \frac{\epsilon}{2}.$$

Similarly, by the definition of the infimum, there exists a step function $\psi \in T[a,b]$ with $\psi \geq f$ and

$$\int_a^b \psi(x)\,dx - U < \frac{\epsilon}{2}.$$

It follows that

$$\begin{aligned}\int_a^b \psi(x)\,dx - \int_a^b \varphi(x)\,dx &= (\int_a^b \psi(x)\,dx - U) + (U - L) \\ &\quad + (L - \int_a^b \varphi(x)\,dx) \\ &< \frac{\epsilon}{2} + 0 + \frac{\epsilon}{2} = \epsilon.\end{aligned}$$

Thus the desired step functions exist.

$\Leftarrow$: Conversely, assume that the stated property holds for every $\epsilon > 0$. Then, for every $\epsilon > 0$, there exist step functions $\varphi, \psi \in T[a,b]$ such that

$$\varphi \leq f \leq \psi \quad \text{and} \quad \int_a^b \psi(x)\,dx - \int_a^b \varphi(x)\,dx < \epsilon.$$

Since $\varphi \leq f$, we have $\int_a^b \varphi(x)\,dx \leq L$. Since $\psi \geq f$, we have $U \leq \int_a^b \psi(x)\,dx$. Hence

$$0 \leq U - L \leq \int_a^b \psi(x)\,dx - \int_a^b \varphi(x)\,dx < \epsilon.$$

Since this holds for every $\epsilon > 0$, it follows that $U - L = 0$, and hence $U = L$. Therefore f is Riemann integrable. $\square$

Theorem 483. *Every continuous function $f : [a, b] \to \mathbb{R}$ is integrable.*

Proof. If $a = b$, then the interval is degenerate and the assertion is trivial. We therefore assume in what follows that $a < b$.

Let $\epsilon > 0$ be given. Since f is continuous on the compact interval $[a, b]$, it is uniformly continuous by the Heine–Cantor theorem. Hence, for

$$\eta := \frac{\epsilon}{b - a}$$

there exists a $\delta > 0$ such that for all $x, y \in [a, b]$ with $|x - y| < \delta$, one has

$$|f(x) - f(y)| < \eta.$$

We now choose a partition

$$a = x_0 < x_1 < \cdots < x_n = b$$

of the interval $[a, b]$ fine enough that

$$x_k - x_{k-1} < \delta \qquad \text{for all } k = 1, \ldots, n.$$

For each subinterval $[x_{k-1}, x_k)$, set

$$m_k := \inf\{f(x) : x \in [x_{k-1}, x_k)\},$$
$$M_k := \sup\{f(x) : x \in [x_{k-1}, x_k)\}.$$

Using these numbers, we define step functions $\varphi, \psi \in T[a, b]$ by

$$\varphi(x) = m_k, \qquad \psi(x) = M_k \quad \text{for } x \in [x_{k-1}, x_k).$$

Then clearly

$$\varphi \leq f \leq \psi.$$

We now show that the integrals of ψ and φ can be made arbitrarily close to each other. Let $u, v \in [x_{k-1}, x_k)$. Then

$$|u - v| \leq x_k - x_{k-1} < \delta,$$

and therefore, by the choice of δ,

$$|f(u) - f(v)| < \eta.$$

Thus the oscillation of f on each subinterval is less than η, and hence

$$M_k - m_k \leq \eta \qquad \text{for all } k = 1, \ldots, n.$$

Consequently, we obtain

$$\begin{aligned}
\int_a^b \psi(x)\,dx - \int_a^b \varphi(x)\,dx &= \sum_{k=1}^{n} (M_k - m_k)(x_k - x_{k-1}) \\
&\leq \sum_{k=1}^{n} \eta\,(x_k - x_{k-1}) \\
&= \eta(b - a) \\
&= \epsilon.
\end{aligned}$$

Thus, for every $\epsilon > 0$, we have found step functions $\varphi, \psi \in T[a, b]$ such that

$$\varphi \leq f \leq \psi \qquad \text{and} \qquad \int_a^b \psi(x)\,dx - \int_a^b \varphi(x)\,dx < \epsilon.$$

By the criterion for Riemann integrability, f is therefore integrable. □

Remark 484. *The proof illustrates the basic idea of Riemann integration particularly clearly: A continuous function varies only slightly on sufficiently small intervals. Therefore, its area can be enclosed arbitrarily accurately from below and from above by step functions.*

Theorem 485. *Every monotone function $f : [a, b] \to \mathbb{R}$ is integrable.*

Proof. We first assume that f is monotonically increasing. The case of a monotonically decreasing function then follows in the same way, with the inequalities reversed.

Thus let f be monotonically increasing, and let $\epsilon > 0$ be given. We choose a partition of the interval $[a, b]$ into n subintervals of equal length:

$$a = x_0 < x_1 < \cdots < x_n = b,$$
$$x_k := a + k\frac{b-a}{n} \quad (k = 0, \ldots, n).$$

On each subinterval $[x_{k-1}, x_k)$, we define step functions by

$$\varphi(x) := f(x_{k-1}), \qquad \psi(x) := f(x_k) \quad \text{for } x \in [x_{k-1}, x_k).$$

Since f is monotonically increasing, for all $x \in [x_{k-1}, x_k)$ we indeed have

$$f(x_{k-1}) \leq f(x) \leq f(x_k),$$

hence

$$\varphi \leq f \leq \psi.$$

Now we compute the difference of the integrals:

$$\begin{aligned}\int_a^b \psi(x)\,dx - \int_a^b \varphi(x)\,dx &= \sum_{k=1}^{n}(f(x_k) - f(x_{k-1}))(x_k - x_{k-1}) \\ &= \frac{b-a}{n}\sum_{k=1}^{n}(f(x_k) - f(x_{k-1})).\end{aligned}$$

The sum can be written out as

$$(f(x_1)-f(x_0))+(f(x_2)-f(x_1))+\cdots+(f(x_n)-f(x_{n-1})).$$

Here each intermediate value $f(x_1), \ldots, f(x_{n-1})$ cancels once with a positive sign and once with a negative sign. Only the first and the last term remain. Therefore,

$$\sum_{k=1}^{n}(f(x_k) - f(x_{k-1})) = f(x_n) - f(x_0) = f(b) - f(a).$$

It follows that

$$\int_a^b \psi(x)\,dx - \int_a^b \varphi(x)\,dx = \frac{b-a}{n}(f(b) - f(a)).$$

Now choose n so large that

$$\frac{b-a}{n}(f(b) - f(a)) < \epsilon,$$

then we obtain step functions $\varphi, \psi \in T[a, b]$ such that

$$\varphi \leq f \leq \psi \qquad \text{and} \qquad \int_a^b \psi(x)\,dx - \int_a^b \varphi(x)\,dx < \epsilon.$$

By the criterion for Riemann integrability, f is therefore integrable.

If f is monotonically decreasing, the argument proceeds analogously: one again chooses a partition into subintervals of equal length and then defines the lower and upper step functions with the endpoint values interchanged. In this case as well, one obtains step functions φ, ψ such that

$$\varphi \leq f \leq \psi \quad \text{and} \quad \int_a^b \psi(x)\,dx - \int_a^b \varphi(x)\,dx < \epsilon.$$

Thus f is integrable in this case as well. □

Remark 486. *A sum of this form is called a* telescoping sum, *because most of its terms cancel each other and only the boundary terms remain.*

Remark 487. *The proof shows that monotonicity already suffices to squeeze the function between two step functions on increasingly fine partitions. Unlike in the case of continuous functions, one does not need small oscillation on every subinterval here; the order property of the function is enough.*

Theorem 488. *Let $f, g : [a, b] \to \mathbb{R}$ be Riemann integrable, and let $\lambda \in \mathbb{R}$. Then $f + g$ and λf are also integrable, and the following statements hold:*

1. $\displaystyle\int_a^b (f+g)(x)\,dx = \int_a^b f(x)\,dx + \int_a^b g(x)\,dx.$

2. $\displaystyle\int_a^b (\lambda f)(x)\,dx = \lambda \int_a^b f(x)\,dx.$

3. *If* $f \leq g$, *then* $\displaystyle\int_a^b f(x)\,dx \leq \int_a^b g(x)\,dx.$

Proof. We use the criterion for Riemann integrability: For every $\epsilon > 0$, there exist suitable step functions that bound the function under consideration from below and from above, and whose integrals differ by less than ϵ.

Proof of 1. Let f and g be integrable. Then, for a given $\epsilon > 0$, there exist step functions

$$\varphi_1, \psi_1, \varphi_2, \psi_2 \in T[a,b]$$

such that

$$\varphi_1 \leq f \leq \psi_1, \qquad \varphi_2 \leq g \leq \psi_2,$$

and

$$\int_a^b \psi_1(x)\,dx - \int_a^b \varphi_1(x)\,dx < \frac{\epsilon}{2},$$
$$\int_a^b \psi_2(x)\,dx - \int_a^b \varphi_2(x)\,dx < \frac{\epsilon}{2}.$$

By the linearity of step functions, it follows that

$$\varphi_1 + \varphi_2 \leq f + g \leq \psi_1 + \psi_2.$$

Moreover,

$$\begin{aligned}\int_a^b (\psi_1+\psi_2)(x)\,dx &- \int_a^b (\varphi_1+\varphi_2)(x)\,dx \\ &= \left(\int_a^b \psi_1(x)\,dx - \int_a^b \varphi_1(x)\,dx\right) \\ &\quad + \left(\int_a^b \psi_2(x)\,dx - \int_a^b \varphi_2(x)\,dx\right) \\ &< \epsilon.\end{aligned}$$

Thus, by the criterion for Riemann integrability, $f+g$ is also integrable.

Applying the same criterion to $f+g$ and using the linearity for step functions, we obtain for the integral:

$$\int_a^b (f+g)(x)\,dx = \int_a^b f(x)\,dx + \int_a^b g(x)\,dx.$$

Proof of 2. First assume that $\lambda \geq 0$. For a given $\epsilon > 0$, choose step functions $\varphi, \psi \in T[a,b]$ such that

$$\varphi \leq f \leq \psi \qquad \text{and} \qquad \int_a^b \psi(x)\,dx - \int_a^b \varphi(x)\,dx < \frac{\epsilon}{\lambda+1}.$$

Then also

$$\lambda\varphi \leq \lambda f \leq \lambda\psi,$$

and, by the linearity for step functions,

$$\begin{aligned}\int_a^b (\lambda\psi)(x)\,dx &- \int_a^b (\lambda\varphi)(x)\,dx \\ &= \lambda\left(\int_a^b \psi(x)\,dx - \int_a^b \varphi(x)\,dx\right).\end{aligned}$$

For $\lambda > 0$, this difference can be made arbitrarily small; for $\lambda = 0$, the function λf is the zero function and hence is integrable anyway.

If $\lambda < 0$, then $-\lambda > 0$. Since $-f$ is also integrable, the case already proved implies that $-\lambda(-f) = \lambda f$ is integrable. Thus λf is integrable in every case.

For the integral itself, it follows that

$$\int_a^b (\lambda f)(x)\,dx = \lambda \int_a^b f(x)\,dx.$$

Proof of 3. Let $f \leq g$. Then

$$h := g - f \geq 0.$$

By the two parts already proved, h is integrable. Since the zero function is a step function and

$$0 \leq h,$$

the function 0 is admissible in the definition of the lower integral of h. Thus the lower integral of h is at least 0. Since h is integrable, the lower and upper integrals agree, and hence

$$\int_a^b h(x)\,dx \geq 0.$$

Using the additivity just proved and compatibility with scalar multiplication, it follows that

$$\int_a^b g(x)\,dx - \int_a^b f(x)\,dx = \int_a^b (g-f)(x)\,dx = \int_a^b h(x)\,dx \geq 0.$$

Therefore,

$$\int_a^b f(x)\,dx \leq \int_a^b g(x)\,dx.$$

This proves all three assertions. □

Theorem 489. *Let $f, g : [a, b] \to \mathbb{R}$ be Riemann integrable. Then the following statements hold:*

1. *The functions*

$$\begin{aligned} f_+(x) &:= \max\{f(x), 0\}, \\ f_-(x) &:= \max\{-f(x), 0\}, \\ |f| & \qquad \text{where } |f| = f_+ + f_-. \end{aligned}$$

are integrable. In particular,

$$f = f_+ - f_-, \qquad |f| = f_+ + f_-.$$

2. *The product fg is integrable.*

3. *For every $n \in \mathbb{N}$, the function $|f|^n$ is also integrable.*

Proof. We use the criterion for Riemann integrability in all parts.

Part 1. Let f be integrable. Then, for every $\epsilon > 0$, there exist step functions

$$\varphi, \psi \in T[a, b] \quad \text{with} \quad \varphi \leq f \leq \psi$$

and
$$\int_a^b \psi(x)\,dx - \int_a^b \varphi(x)\,dx < \epsilon.$$
We now define
$$\varphi_+(x) := \max\{\varphi(x), 0\}, \qquad \psi_+(x) := \max\{\psi(x), 0\}.$$

Since φ and ψ are step functions, φ_+ and ψ_+ are also step functions. Moreover, pointwise we have
$$\varphi_+ \leq f_+ \leq \psi_+.$$

Indeed, applying the increasing function $t \mapsto \max\{t, 0\}$ to $\varphi \leq f \leq \psi$ gives exactly this estimate.

Furthermore, for all $x \in [a, b]$,
$$0 \leq \psi_+(x) - \varphi_+(x) \leq \psi(x) - \varphi(x).$$

Indeed, the positive-part function $t \mapsto t_+$ cannot increase the distance between two numbers. It follows that
$$\begin{aligned} 0 &\leq \int_a^b \psi_+(x)\,dx - \int_a^b \varphi_+(x)\,dx \\ &\leq \int_a^b \psi(x)\,dx - \int_a^b \varphi(x)\,dx \\ &< \epsilon. \end{aligned}$$

Hence, by the criterion for Riemann integrability, f_+ is integrable.

Applying the same argument to $-f$, we see that
$$f_- = (-f)_+$$

is also integrable. Therefore,

$$|f| = f_+ + f_-$$

is integrable as well, since the sum of integrable functions is integrable.

Moreover, the identities

$$f = f_+ - f_-, \qquad |f| = f_+ + f_-$$

follow directly from the definitions of f_+ and f_-.

Part 2. Since f and g are integrable, they are bounded. Hence there exist numbers $M, N \geq 0$ such that

$$|f(x)| \leq M, \qquad |g(x)| \leq N \qquad \text{for all } x \in [a, b].$$

Let $\epsilon > 0$ be given. Since f and g are integrable, there exist step functions

$$\varphi, \psi, \eta, \theta \in T[a, b]$$

with

$$\varphi \leq f \leq \psi, \qquad \eta \leq g \leq \theta,$$

and

$$\int_a^b \psi(x)\,dx - \int_a^b \varphi(x)\,dx \leq \frac{\epsilon}{2(M+N)+1},$$

$$\int_a^b \theta(x)\,dx - \int_a^b \eta(x)\,dx < \frac{\epsilon}{2(M+N)+1}.$$

We now consider a common refinement of the partitions on which $\varphi, \psi, \eta, \theta$ are respectively constant. On each subinterval $[x_{k-1}, x_k)$, let these constant values be

$$\varphi = \alpha_k, \qquad \psi = \beta_k, \qquad \eta = \gamma_k, \qquad \theta = \delta_k.$$

Then, for every $x \in [x_{k-1}, x_k)$,

$$\alpha_k \leq f(x) \leq \beta_k, \qquad \gamma_k \leq g(x) \leq \delta_k.$$

Thus the four possible endpoint values of the product are

$$\alpha_k\gamma_k,\ \alpha_k\delta_k,\ \beta_k\gamma_k,\ \beta_k\delta_k.$$

We set

$$p_k := \min\{\alpha_k\gamma_k, \alpha_k\delta_k, \beta_k\gamma_k, \beta_k\delta_k\},$$

$$q_k := \max\{\alpha_k\gamma_k, \alpha_k\delta_k, \beta_k\gamma_k, \beta_k\delta_k\}.$$

Then, on this subinterval,

$$p_k \leq f(x)g(x) \leq q_k.$$

We now estimate the oscillation. For any two endpoint products, we have

$$u_1v_1 - u_2v_2 = u_1(v_1 - v_2) + v_2(u_1 - u_2).$$

Hence, using $|u_1| \leq M$ and $|v_2| \leq N$, it follows that

$$|u_1v_1 - u_2v_2| \leq M|v_1 - v_2| + N|u_1 - u_2|.$$

In particular, the difference between the largest and the smallest of the four endpoint values is at most

$$M(\delta_k - \gamma_k) + N(\beta_k - \alpha_k).$$

Thus

$$q_k - p_k \le M(\delta_k - \gamma_k) + N(\beta_k - \alpha_k).$$

If we now define step functions $p, q \in T[a, b]$ by

$$p(x) = p_k, \qquad q(x) = q_k \quad \text{for } x \in [x_{k-1}, x_k),$$

then it follows that

$$p \le fg \le q.$$

Moreover,

$$\begin{aligned}
\int_a^b q(x)\,dx - \int_a^b p(x)\,dx &= \sum_{k=1}^n (q_k - p_k)(x_k - x_{k-1}) \\
&\le M \sum_{k=1}^n (\delta_k - \gamma_k)(x_k - x_{k-1}) \\
&\quad + N \sum_{k=1}^n (\beta_k - \alpha_k)(x_k - x_{k-1}) \\
&= M \left(\int_a^b \theta(x)\,dx - \int_a^b \eta(x)\,dx \right) \\
&\quad + N \left(\int_a^b \psi(x)\,dx - \int_a^b \varphi(x)\,dx \right) \\
&< \epsilon.
\end{aligned}$$

Hence, by the criterion for Riemann integrability, fg is integrable as well.

Part 3. By Part 1, $|f|$ is integrable and $|f| \geq 0$. For $n \in \mathbb{N}$, the integrability of $|f|^n$ now follows by repeated application of Part 2:

$$|f|^n = \underbrace{|f| \cdot |f| \cdots |f|}_{n \text{ times}}.$$

Thus $|f|^n$ is integrable. □

Remark 490. *For general real exponents $a \geq 1$, the integrability of $|f|^a$ is a very natural next step, but it is most elegantly proved by a later theorem on the integrability of continuous functions under composition.*

Remark 491. *In general, it is not true that*

$$\int_a^b f(x)g(x)\,dx = \left(\int_a^b f(x)\,dx\right)\left(\int_a^b g(x)\,dx\right).$$

A counterexample is already obtained on the interval $[0,1]$ by taking

$$f(x) = g(x) = x.$$

Then

$$\int_0^1 f(x)g(x)\,dx = \int_0^1 x^2\,dx = \frac{1}{3},$$

whereas

$$\left(\int_0^1 f(x)\,dx\right)\left(\int_0^1 g(x)\,dx\right) = \left(\int_0^1 x\,dx\right)^2 = \left(\frac{1}{2}\right)^2 = \frac{1}{4}.$$

Thus the two sides are different.

14.2.3 Mean Value Theorem for Integrals

Theorem 492 (Mean Value Theorem for Integrals). *Let $f, g : [a, b] \to \mathbb{R}$ be continuous, and let $g \geq 0$. Then there exists a $c \in [a, b]$ such that*

$$\int_a^b f(x)g(x)\,dx = f(c)\int_a^b g(x)\,dx.$$

Proof. Since f is continuous on the compact interval $[a, b]$, the maximum–minimum theorem implies that f attains a minimum and a maximum there. Thus there exist numbers $m, M \in \mathbb{R}$ such that

$$m \leq f(x) \leq M \qquad \text{for all } x \in [a, b].$$

Since, moreover, $g(x) \geq 0$ for all $x \in [a, b]$, we obtain pointwise

$$m\,g(x) \leq f(x)g(x) \leq M\,g(x).$$

All three functions are continuous and therefore integrable. By the monotonicity of the integral, it follows that

$$m\int_a^b g(x)\,dx \leq \int_a^b f(x)g(x)\,dx \leq M\int_a^b g(x)\,dx.$$

Set

$$G := \int_a^b g(x)\,dx, \qquad I := \int_a^b f(x)g(x)\,dx,$$

so that this estimate becomes

$$mG \leq I \leq MG.$$

First consider the case $G = 0$. Since $g \geq 0$ and g is continuous, this implies that $g(x) = 0$ for all $x \in [a, b]$, and hence also $I = 0$. Thus the assertion holds for every $c \in [a, b]$.

Now suppose that $G > 0$. Then we may divide by G and obtain

$$m \leq \frac{I}{G} \leq M.$$

Thus the number $\frac{I}{G}$ lies between the minimum and the maximum of f. Since f is continuous, the intermediate value theorem implies that f assumes every value between m and M. Hence there exists a $c \in [a, b]$ such that

$$f(c) = \frac{I}{G}.$$

It follows that

$$\int_a^b f(x)g(x)\,dx = I = f(c)\,G = f(c)\int_a^b g(x)\,dx.$$

This is exactly what had to be shown. □

Remark 493. *Intuitively, the theorem says that the integral of f against a nonnegative weight function g is a weighted mean value of f. Therefore, the value of the integral lies between the minimum and the maximum of f, and by continuity it is actually attained at some point c.*

Theorem 494 (Simple mean value theorem for integrals). *If $f : [a, b] \to \mathbb{R}$ is continuous, then there exists a $c \in [a, b]$ such that*

$$\int_a^b f(x)\,dx = f(c)(b-a).$$

Proof. We apply the mean value theorem for integrals from the preceding theorem to the constant function

$$g(x) := 1 \qquad (x \in [a, b])$$

Then g is continuous and nonnegative, and we have

$$\int_a^b f(x)g(x)\,dx = \int_a^b f(x)\,dx$$

as well as

$$\int_a^b g(x)\,dx = \int_a^b 1\,dx = b-a.$$

Hence, by the mean value theorem for integrals, there exists a $c \in [a, b]$ such that

$$\int_a^b f(x)\,dx = f(c)\int_a^b 1\,dx = f(c)(b-a).$$

□

Remark 495. *Intuitively, the theorem says that the integral value of a continuous function is equal to the function value at a suitable point multiplied by the length of the interval. Thus, the integral measures an average function value.*

So far, we have described the integral by means of step functions as well as upper and lower integrals, and we have studied important properties of integrable functions. Throughout, the underlying idea has been to approximate the area under a graph by rectangles.

This idea leads directly to *Riemann sums*. One partitions the interval $[a, b]$ into many small subintervals, replaces the function on each subinterval by a suitable function value, and adds the resulting rectangular areas. As the subintervals become finer and finer, these sums approach the integral.

Riemann sums therefore provide a particularly intuitive description of the integral and, at the same time, establish the connection between the geometric idea of area and the analytical definition of the integral.

14.3 Riemann Sums

The definition of the integral by means of step functions and upper and lower integrals is mathematically very precise. Nevertheless, the original geometric intuition remains central: the area under the graph of a function is to be approximated by many narrow rectangles.

This is exactly what leads to Riemann sums. One partitions the interval $[a, b]$ into small subintervals, chooses a sample point in each subinterval, and replaces the function there by the corresponding rectangle height. The sum of the rectangular areas then gives an approximation of the integral. The finer the partition is, the more accurate the approximation becomes.

14.3.1 Partitions and Mesh Size

Definition 496 (Partition). *A* partition *of the interval $[a, b]$ is a finite sequence of points*

$$a = x_0 < x_1 < \cdots < x_n = b.$$

The corresponding subintervals are

$$[x_{k-1}, x_k] \qquad (k = 1, \ldots, n).$$

Definition 497 (Mesh size). *The* mesh size *of a partition*

$$Z: \ a = x_0 < x_1 < \cdots < x_n = b$$

is defined by

$$|Z| := \max_{1 \le k \le n} (x_k - x_{k-1}).$$

The smaller $|Z|$ is, the finer the partition is.

Definition 498 (Riemann sum). *Let $f : [a, b] \to \mathbb{R}$ be a function and let*

$$Z: \ a = x_0 < x_1 < \cdots < x_n = b$$

be a partition. If, for each subinterval $[x_{k-1}, x_k]$, one chooses a sample point

$$\xi_k \in [x_{k-1}, x_k],$$

then

$$S(f, Z, \xi) := \sum_{k=1}^{n} f(\xi_k)\,(x_k - x_{k-1})$$

is called the corresponding Riemann sum.

Intuitively, $f(\xi_k)\,(x_k - x_{k-1})$ is the area of a rectangle with width $x_k - x_{k-1}$ and height $f(\xi_k)$. Thus the entire sum describes the area under the graph of f by means of a rectangular approximation.

14.3.2 Riemann Sums and the Integral

Theorem 499 (Convergence of Riemann sums)**.** *Let $f : [a, b] \to \mathbb{R}$ be Riemann integrable. Then, for every sequence of partitions Z_n with*

$$|Z_n| \to 0$$

and for every choice of sample points $\xi_k^{(n)} \in [x_{k-1}^{(n)}, x_k^{(n)}]$, one has

$$S(f, Z_n, \xi^{(n)}) \to \int_a^b f(x)\,dx.$$

Proof. Let

$$I := \int_a^b f(x)\,dx.$$

Since f is Riemann integrable, f is in particular bounded. Hence there exists a number $M > 0$ such that

$$|f(x)| \leq M \qquad \text{for all } x \in [a, b].$$

Now let $\epsilon > 0$ be given. By Riemann integrability, there exists a partition

$$P: \ a = x_0 < x_1 < \cdots < x_r = b$$

such that

$$U(f, P) - L(f, P) < \frac{\epsilon}{2},$$

where $L(f, P)$ and $U(f, P)$ denote the lower and upper sums corresponding to P.

Set

$$\eta := \min_{1 \le i \le r} (x_i - x_{i-1}) > 0.$$

We now choose $\delta > 0$ so small that

$$2M(r-1)\delta < \frac{\epsilon}{2}.$$

Let $Z: \ a = t_0 < t_1 < \cdots < t_m = b$ be a partition with

$$|Z| < \delta,$$

and let $\xi_k \in [t_{k-1}, t_k]$ be arbitrary sample points.

We call a subinterval $[t_{k-1}, t_k]$ *good* if it contains none of the points

$$x_1, \ldots, x_{r-1}$$

and a *boundary interval* if it contains at least one of these points. Since $|Z| < \eta$, each boundary interval can contain at most one of these points. Therefore, the total length of all boundary intervals is at most $(r-1)\delta$.

Writing

$$m_i := \inf\{f(x) : x \in [x_{i-1}, x_i]\},$$
$$M_i := \sup\{f(x) : x \in [x_{i-1}, x_i]\}$$

we have, on every good subinterval lying inside $[x_{i-1}, x_i]$,

$$m_i \leq f(\xi_k) \leq M_i.$$

Hence the contribution of these good subintervals lies between the corresponding lower and upper sums associated with P. The boundary intervals can only be estimated roughly, but on them we have, because $|f| \leq M$,

$$|f(\xi_k)(t_k - t_{k-1})| \leq M(t_k - t_{k-1}).$$

Thus the Riemann sum differs from the lower and upper sums controlled by P by at most

$$2M(r-1)\delta$$

in the boundary contribution.

Together with

$$U(f, P) - L(f, P) < \frac{\epsilon}{2}$$

it follows that

$$|S(f, Z, \xi) - I| < \epsilon.$$

Since $\epsilon > 0$ was arbitrary, the Riemann sums converge to the integral. □

Remark 500. *For uniform partitions*

$$x_k = a + k\frac{b-a}{n}$$

one obtains the particularly familiar form

$$\sum_{k=1}^{n} f(\xi_k)\,\frac{b-a}{n}.$$

If the sample points are chosen, for instance, as left endpoints, right endpoints, or midpoints, then the classical rectangle rules arise.

14.3.3 Riemann Sums and Decomposition of the Interval

Example 501 (A simple example of a Riemann sum). *We consider the function $f(x) = x$ on $[0,1]$ and the uniform partition*

$$0 = x_0 < x_1 < \cdots < x_n = 1, \qquad x_k = \frac{k}{n}.$$

If in each subinterval the right endpoint is chosen as the sample point, that is,

$$\xi_k = x_k = \frac{k}{n},$$

then one obtains the Riemann sum

$$S_n = \sum_{k=1}^{n} f(\xi_k)\,(x_k - x_{k-1}) = \sum_{k=1}^{n} \frac{k}{n}\cdot\frac{1}{n} = \frac{1}{n^2}\sum_{k=1}^{n} k = \frac{n+1}{2n}.$$

Thus one sees immediately that these sums converge to $\frac{1}{2}$ *as* $n \to \infty$.

Theorem 502 (Decomposition of an integral). *Let* $a < b < c$ *and* $f : [a, c] \to \mathbb{R}$. *Then* f *is Riemann integrable if and only if the restrictions*

$$f|_{[a,b]} \quad \textit{and} \quad f|_{[b,c]}$$

are Riemann integrable. In this case,

$$\int_a^c f(x)\,dx = \int_a^b f(x)\,dx + \int_b^c f(x)\,dx.$$

Proof. We first prove the two directions of the equivalence.

1. First direction: If f is integrable on $[a, c]$, then its restrictions to $[a, b]$ and $[b, c]$ are also integrable.

Let $\epsilon > 0$ be given. By the criterion for Riemann integrability, there are step functions $\varphi, \psi \in T[a, c]$ such that

$$\varphi \leq f \leq \psi$$

and

$$\int_a^c \psi(x)\,dx - \int_a^c \varphi(x)\,dx < \epsilon.$$

Then the restrictions $\varphi|_{[a,b]}$ and $\psi|_{[a,b]}$ are step functions on $[a, b]$, and

$$\varphi|_{[a,b]} \leq f|_{[a,b]} \leq \psi|_{[a,b]}.$$

Moreover,

$$\int_a^b \psi(x)\,dx - \int_a^b \varphi(x)\,dx \le \int_a^c \psi(x)\,dx - \int_a^c \varphi(x)\,dx < \epsilon.$$

Thus, by the criterion for Riemann integrability, $f|_{[a,b]}$ is integrable. The integrability of $f|_{[b,c]}$ follows in exactly the same way.

2. Second direction: If the restrictions to $[a,b]$ and $[b,c]$ are integrable, then f is integrable on $[a,c]$.

Let $\epsilon > 0$ be given. Then there are step functions $\varphi_1, \psi_1 \in T[a,b]$ and $\varphi_2, \psi_2 \in T[b,c]$ such that

$$\varphi_1 \le f \le \psi_1 \quad \text{on } [a,b], \qquad \varphi_2 \le f \le \psi_2 \quad \text{on } [b,c],$$

and

$$\int_a^b \psi_1(x)\,dx - \int_a^b \varphi_1(x)\,dx \le \frac{\epsilon}{2},$$
$$\int_b^c \psi_2(x)\,dx - \int_b^c \varphi_2(x)\,dx \le \frac{\epsilon}{2}$$

We now define step functions $\varphi, \psi \in T[a,c]$ by

$$\varphi(x) = \begin{cases} \varphi_1(x), & x \in [a,b), \\ \varphi_2(x), & x \in [b,c], \end{cases} \qquad \psi(x) = \begin{cases} \psi_1(x), & x \in [a,b), \\ \psi_2(x), & x \in [b,c]. \end{cases}$$

Then, on all of $[a,c]$, we have

$$\varphi \le f \le \psi.$$

By the additivity of the integral for step functions, it follows that

$$\begin{aligned}\int_a^c \psi(x)\,dx - \int_a^c \varphi(x)\,dx &= \left(\int_a^b \psi_1(x)\,dx - \int_a^b \varphi_1(x)\,dx\right) \\ &\quad + \left(\int_b^c \psi_2(x)\,dx - \int_b^c \varphi_2(x)\,dx\right) \\ &< \frac{\epsilon}{2} + \frac{\epsilon}{2} = \epsilon.\end{aligned}$$

Hence, by the criterion for Riemann integrability, f is integrable on $[a, c]$.

3. Additivity of the integral. Now let $f|_{[a,b]}$ and $f|_{[b,c]}$ be integrable. We write

$$I_{ab} := \int_a^b f(x)\,dx,$$

$$I_{bc} := \int_b^c f(x)\,dx,$$

$$I_{ac} := \int_a^c f(x)\,dx.$$

Let $\epsilon > 0$. As above, we choose step functions $\varphi, \psi \in T[a, c]$ such that

$$\varphi \leq f \leq \psi \qquad \text{and} \qquad \int_a^c \psi(x)\,dx - \int_a^c \varphi(x)\,dx < \epsilon.$$

Then, since $\varphi \leq f \leq \psi$, the monotonicity of the integral for integrable functions also gives

$$\int_a^b \varphi(x)\,dx \le I_{ab} \le \int_a^b \psi(x)\,dx,$$

$$\int_b^c \varphi(x)\,dx \le I_{bc} \le \int_b^c \psi(x)\,dx.$$

Adding these two estimates yields

$$\int_a^c \varphi(x)\,dx \le I_{ab} + I_{bc} \le \int_a^c \psi(x)\,dx,$$

because, for step functions, we already have

$$\int_a^c \varphi(x)\,dx = \int_a^b \varphi(x)\,dx + \int_b^c \varphi(x)\,dx$$

and likewise for ψ.

The same is true for I_{ac}, hence

$$\int_a^c \varphi(x)\,dx \le I_{ac} \le \int_a^c \psi(x)\,dx.$$

Thus I_{ac} and $I_{ab} + I_{bc}$ both lie in an interval of length

$$\int_a^c \psi(x)\,dx - \int_a^c \varphi(x)\,dx < \epsilon.$$

Consequently,

$$|I_{ac} - (I_{ab} + I_{bc})| < \epsilon.$$

Since $\epsilon > 0$ was arbitrary, it follows that

$$I_{ac} = I_{ab} + I_{bc}.$$

This proves the claim. □

The preceding theorem shows that integrals over adjacent subintervals behave additively. For example, if the interval $[a, x]$ is split at a point y with $a \leq y \leq x$, then

$$\int_a^x f(t)\,dt = \int_a^y f(t)\,dt + \int_y^x f(t)\,dt.$$

Thus, in a certain sense, the integral behaves like an accumulated quantity: if the interval of integration is enlarged, only the contribution from the newly added part is added.

This observation leads to an important idea. Up to now, the limits of integration have been fixed. We now regard the upper limit of integration as a variable quantity. Thus, for an integrable function f, we define the function

$$F(x) := \int_a^x f(t)\,dt.$$

It assigns to each point $x \in [a, b]$ the area accumulated up to that point under the graph of f.

This so-called *integral function* will turn out to be the decisive link between integration and differentiation and leads directly to the Fundamental Theorem of Calculus.

Chapter 15

The Fundamental Theorem of Calculus

In the preceding chapters, we have encountered two central concepts of analysis: the derivative as a local rate of change and the integral as a notion of area or summation. At first, these two concepts seemed quite different. The derivative describes the behavior of a function in a small neighborhood of a point, whereas the integral combines many small contributions into one total value.

In this chapter, however, we shall see that there is a deep connection between these two concepts. Under suitable assumptions, integration can be used to construct from a function a new function whose derivative is precisely the original function. In this sense, integration turns out to be a kind of inverse operation to differentiation. This connection is one of the most important results in analysis, since it not only explains the theoretical relationship between derivatives and integrals, but also provides a fundamental tool for computing integrals.

The central question is how integrals can be determined efficiently. Instead of evaluating every integral directly by means of Riemann sums or limits, we shall later see that many integrals can be computed with the help of antiderivatives. The Fundamental Theorem of Calculus therefore combines two initially separate areas into one unified whole. It consists of two closely related statements, which we shall formulate and prove separately.

15.1 Integral Functions

So far, when considering integrals, we have always worked with two fixed limits of integration. For a given function, the integral then produced a single numerical value. We now allow one of the limits to vary. In this way, the integral itself gives rise to a function.

Let $f : I \to \mathbb{R}$ be a continuous function on an interval $I \subset \mathbb{R}$, and let $a \in I$ be a fixed point. For each $x \in I$, we define

$$F(x) := \int_a^x f(t)\,dt.$$

The mapping $F : I \to \mathbb{R}$ is called the *integral function* of f with lower limit a. The idea behind this is simple:

to each point x, we assign the area, or more precisely the integral of f, between a and x. Thus the value of the integral no longer depends only on the function f, but also on the position of the right endpoint x.

The following theorem shows that this new function is very regular: it is differentiable, and its derivative is precisely the original function f.

Theorem 503 (Fundamental Theorem of Calculus, Part 1). *Let $f : I \to \mathbb{R}$ be a continuous function and let $a \in I$. For $x \in I$, let*

$$F(x) := \int_a^x f(t)\,dt.$$

Then the function $F : I \to \mathbb{R}$ is differentiable, and

$$F'(x) = f(x) \qquad \text{for all } x \in I.$$

This is the first part of the Fundamental Theorem of Calculus.

Proof. First, let $x \in I$ be arbitrary and let $h \neq 0$ be such that $x + h \in I$. Then

$$\begin{aligned}\frac{F(x+h) - F(x)}{h} &= \frac{1}{h}\left(\int_a^{x+h} f(t)\,dt - \int_a^x f(t)\,dt\right)\\ &= \frac{1}{h}\int_x^{x+h} f(t)\,dt.\end{aligned}$$

We now apply the mean value theorem for integrals to the continuous function f on the interval between x and

$x+h$. Hence there exists a point ξ_h between x and $x+h$ such that

$$\int_x^{x+h} f(t)\,dt = f(\xi_h)\,h.$$

Therefore,

$$\frac{F(x+h)-F(x)}{h} = f(\xi_h).$$

Since ξ_h lies between x and $x+h$, we have $\xi_h \to x$ as $h \to 0$. By the continuity of f at x, it follows that

$$\lim_{h\to 0} \frac{F(x+h)-F(x)}{h} = \lim_{h\to 0} f(\xi_h) = f(x).$$

Hence F is differentiable at x and $F'(x) = f(x)$. Since $x \in I$ was arbitrary, this holds for all $x \in I$. □

Remark 504. *The theorem shows that integrating a continuous function f produces a differentiable function F whose derivative is again f. Thus integration and differentiation are closely connected. This connection is precisely what will be used in the remainder of the chapter to compute integrals by means of antiderivatives.*

15.1.1 Antiderivatives

The first part of the fundamental theorem shows that, starting from a continuous function f, integration produces a differentiable function whose derivative is again f. Such functions play a central role and are given a name of their own.

Definition 505. *Let $f : I \to \mathbb{R}$ be a function on an interval $I \subset \mathbb{R}$. A function $F : I \to \mathbb{R}$ is called an* antiderivative *of f if F is differentiable and, for all $x \in I$,*

$$F'(x) = f(x).$$

Antiderivatives are therefore precisely those functions whose derivative is the given function f. It is important to note that, in general, a function does not have only one antiderivative: if one adds a constant to an antiderivative, one obtains another antiderivative. The following result makes this relationship precise.

Proposition 506. *Let $F, G : I \to \mathbb{R}$ be antiderivatives of $f : I \to \mathbb{R}$. Then:*

$$G \text{ is also an antiderivative of } f$$

$$\iff F - G \text{ is constant.}$$

Proof. First assume that F and G are antiderivatives of f. Then, for all $x \in I$,

$$(F - G)'(x) = F'(x) - G'(x) = f(x) - f(x) = 0.$$

Thus the function $F - G$ has derivative 0 everywhere. Since I is an interval, it follows from the mean value theorem that $F - G$ is constant.

Conversely, suppose that $F - G$ is constant. Then there is a constant $c \in \mathbb{R}$ such that

$$F(x) - G(x) = c \qquad \text{for all } x \in I.$$

It follows that

$$G(x) = F(x) - c.$$

Differentiating gives

$$G'(x) = F'(x) = f(x) \qquad \text{for all } x \in I.$$

Hence G is also an antiderivative of f. □

Remark 507. *The proposition shows that antiderivatives differ at most by a constant. For this reason, one often speaks not of* the *antiderivative, but of* an *antiderivative or of the general antiderivative*

$$F(x) + c, \qquad c \in \mathbb{R}.$$

Theorem 508 (Fundamental theorem of calculus, second part)**.** *Let $f : I \to \mathbb{R}$ be a continuous function, and let F be an antiderivative of f. Then, for all $a, b \in I$, one has*

$$\int_a^b f(x)\,dx = F(b) - F(a).$$

Proof. Fix an arbitrary but fixed $a \in I$ and define the integral function

$$G(x) := \int_a^x f(t)\,dt \qquad (x \in I).$$

By the first part of the fundamental theorem of calculus, G is an antiderivative of f. Since F is also an antiderivative of f, the preceding proposition shows that the difference $G - F$ is constant. Hence there exists a constant $c \in \mathbb{R}$ such that

$$G(x) - F(x) = c \qquad \text{for all } x \in I.$$

Setting $x = a$, we obtain

$$G(a) - F(a) = c.$$

Since

$$G(a) = \int_a^a f(t)\,dt = 0$$

it follows that

$$c = -F(a).$$

Thus, for all $x \in I$, we have

$$G(x) = F(x) - F(a).$$

In particular, for $x = b$ we obtain

$$\int_a^b f(t)\,dt = G(b) = F(b) - F(a).$$

This proves the claim. □

This theorem is of central importance, since it shows that definite integrals can be computed not by direct summation, but by means of suitable antiderivatives.

15.2 Examples for Computing Definite Integrals

The second part of the fundamental theorem of calculus shows that definite integrals can be computed by means of antiderivatives. The following simple examples illustrate how integration rules follow directly from differentiation rules.

Example 509. *We have*

$$\int_a^b x^n\,dx = \left[\frac{x^{n+1}}{n+1}\right]_a^b = \frac{b^{n+1}-a^{n+1}}{n+1} \qquad (n \neq -1).$$

Example 510. *Since* $(\sin x)' = \cos x$*, the fundamental theorem gives*

$$\int_a^b \cos x\,dx = \sin(b) - \sin(a).$$

Similarly, since $(e^x)' = e^x$*, we obtain*

$$\int_a^b e^x\,dx = e^b - e^a.$$

15.3 Summary: The Fundamental Theorem of Calculus

We have now seen that differentiation and integration are deeply connected. The fundamental theorem of calculus makes precise that, under suitable assumptions, these two operations undo each other: By integrating a continuous function, one obtains a differentiable function whose derivative is the original function. Conversely, definite integrals can be computed with the help of antiderivatives.

Theorem 511 (Fundamental Theorem of Calculus). *Let $f : I \to \mathbb{R}$ be continuous, and let $a \in I$. Then the integral function*

$$F(x) = \int_a^x f(t)\,dt$$

is differentiable and satisfies

$$F'(x) = f(x) \qquad \text{for all } x \in I.$$

Conversely, if G is an antiderivative of f, then for all $a, b \in I$,

$$\int_a^b f(x)\,dx = G(b) - G(a).$$

Thus, the fundamental theorem shows that integration can be understood not only as a limiting process of Riemann sums, but also as the inverse operation to differentiation. It is precisely this connection that makes the integral an extremely powerful tool for computing areas, total changes, and many other quantities in analysis.

15.4 Methods of Integration

The fundamental theorem of calculus shows that the computation of definite integrals can be reduced to the determination of antiderivatives. In many cases, however, an antiderivative is not immediately apparent. We therefore need systematic methods that help us carry out suitable transformations.

The first and most important of these methods are based on the differentiation rules already known. The substitution rule is, in a sense, the inverse of the chain rule: if an integrand consists of a composite function and the derivative of the inner function, then a suitable change of variables can simplify the integral. This connection is described precisely in the following theorem.

15.4.1 The Substitution Rule

Theorem 512 (Substitution rule). *Let $f : I \to \mathbb{R}$ be a continuous function and let*

$$\phi : [a, b] \to \mathbb{R}$$

be a continuously differentiable function such that

$$\phi([a, b]) \subset I.$$

Then

$$\int_a^b f(\phi(t))\, \phi'(t)\, dt = \int_{\phi(a)}^{\phi(b)} f(x)\, dx.$$

Proof. Since f is continuous on the interval I, the function f has an antiderivative $F : I \to \mathbb{R}$. Thus

$$F'(x) = f(x) \qquad \text{for all } x \in I.$$

Now consider the composite function $F \circ \phi$. By the chain rule, it is differentiable and satisfies

$$(F \circ \phi)'(t) = F'(\phi(t)) \cdot \phi'(t) = f(\phi(t))\, \phi'(t) \qquad \text{for all } t \in [a, b]$$

Hence $F \circ \phi$ is an antiderivative of the function $t \mapsto f(\phi(t))\, \phi'(t)$.

By the second part of the fundamental theorem of calculus, it follows that

$$\int_a^b f(\phi(t))\, \phi'(t)\, dt = (F \circ \phi)(b) - (F \circ \phi)(a).$$

Therefore

$$\int_a^b f(\phi(t))\, \phi'(t)\, dt = F(\phi(b)) - F(\phi(a)).$$

Applying the second part of the fundamental theorem once more to the function f, we obtain

$$F(\phi(b)) - F(\phi(a)) = \int_{\phi(a)}^{\phi(b)} f(x)\, dx.$$

This proves the assertion. □

Remark 513. *The substitution rule is the integral form of the chain rule. In practice, it is often read as the instruction:* Recognize an inner function, introduce it as a new variable, and at the same time replace its derivative in the differential.

15.4.2 Examples of the Substitution Rule

We illustrate the substitution rule with a few simple examples. In each case, one should pay attention to whether the integrand contains an inner function together with its derivative.

Example 514. *Let*

$$I = \int_0^1 2x\cos(x^2)\,dx.$$

We set

$$u = x^2.$$

Then

$$du = 2x\,dx,$$

and the limits of integration become

$$u(0) = 0, \qquad u(1) = 1.$$

Thus we obtain

$$I = \int_0^1 \cos(u)\,du = \sin(u)\Big|_0^1 = \sin(1) - \sin(0) = \sin(1).$$

Example 515. *Let*

$$J = \int_0^1 \frac{2x}{1+x^2}\,dx.$$

With the substitution

$$u = 1 + x^2 \qquad \textit{and} \qquad du = 2x\,dx$$

it follows that

$$J = \int_1^2 \frac{1}{u}\,du = \ln(u)\Big|_1^2 = \ln(2).$$

Example 516. *For*

$$K = \int_0^{\pi} \sin(x)\cos(x)\,dx$$

we use the substitution

$$u = \sin(x), \qquad du = \cos(x)\,dx.$$

Since $\sin(0) = 0$ *and* $\sin(\pi) = 0$, *this becomes*

$$K = \int_0^0 u\,du = 0.$$

These examples show that the substitution rule is especially useful when the integrand has the form of a composite function together with the derivative of the inner function.

15.4.3 Integration by Parts

Not every integral can be simplified directly by substitution. Often, one instead encounters a product of two functions, where one of the two functions becomes simpler when differentiated. In precisely such cases, it is useful to rewrite the integral in such a way that the derivative is, in a sense, "transferred" from one factor to the other. This leads to the formula for integration by parts.

The idea behind it is simple: Since the product rule for differentiation describes how to differentiate a product, it can be translated into an integration rule by means of the fundamental theorem of calculus. Thus, integration by parts is, in a certain sense, the reverse of the product rule.

Theorem 517 (Integration by Parts). *For two continuously differentiable functions $f, g : [a, b] \to \mathbb{R}$, one has*

$$\int_a^b f(x)\, g'(x)\, dx = f(x)g(x)|_a^b - \int_a^b f'(x)\, g(x)\, dx.$$

Equivalently, one also writes

$$\int_a^b f\, dg = fg - \int_a^b g\, df.$$

Proof. We set

$$F(x) := f(x)g(x).$$

Then F is differentiable by the product rule, and

$$F'(x) = f'(x)g(x) + f(x)g'(x).$$

We now integrate this equation over the interval $[a, b]$:

$$\int_a^b F'(x)\, dx = \int_a^b f'(x)g(x)\, dx + \int_a^b f(x)g'(x)\, dx.$$

By the second part of the fundamental theorem of calculus,

$$\int_a^b F'(x)\, dx = F(b) - F(a) = f(x)g(x)|_a^b .$$

Hence,

$$\int_a^b f'(x)g(x)\,dx + \int_a^b f(x)g'(x)\,dx = f(x)g(x)\big|_a^b.$$

Rearranging gives

$$\int_a^b f(x)\,g'(x)\,dx = f(x)g(x)\big|_a^b - \int_a^b f'(x)\,g(x)\,dx.$$

This proves the claim. □

Remark 518. *Integration by parts is especially useful when one of the two factors becomes simpler upon differentiation. Typical examples are products such as* xe^x, $x\sin x$, *or* $x\ln x$.

Examples

Example 519. *We compute*

$$\int xe^x\,dx.$$

For this purpose, we choose

$$f(x) = x, \qquad g'(x) = e^x.$$

Then

$$f'(x) = 1, \qquad g(x) = e^x,$$

and integration by parts gives

$$\int xe^x\,dx = xe^x - \int e^x\,dx = xe^x - e^x + C.$$

Thus we obtain

$$\int xe^x \, dx = e^x(x-1) + C.$$

Example 520. *For the integral*

$$\int x \cos x \, dx$$

we set

$$f(x) = x, \qquad g'(x) = \cos x.$$

Then

$$f'(x) = 1, \qquad g(x) = \sin x.$$

Hence we obtain

$$\int x \cos x \, dx = x \sin x - \int \sin x \, dx = x \sin x + \cos x + C.$$

Example 521. *We consider*

$$\int_1^e \ln(x) \, dx.$$

Here we choose

$$f(x) = \ln(x), \qquad g'(x) = 1.$$

Then

$$f'(x) = \frac{1}{x}, \qquad g(x) = x.$$

Therefore,

$$\int_1^e \ln(x) \, dx = x \ln(x)|_1^e - \int_1^e 1 \, dx.$$

Since

$$x\ln(x)|_1^e = e\cdot 1 - 1\cdot 0 = e \quad \textit{and} \quad \int_1^e 1\,dx = e-1,$$

we get

$$\int_1^e \ln(x)\,dx = e-(e-1) = 1.$$

Example 522. *Integration by parts can also be very useful for definite integrals. For example, consider*

$$\int_0^1 x\ln(x)\,dx.$$

We set

$$f(x) = \ln(x), \qquad g'(x) = x.$$

Then

$$f'(x) = \frac{1}{x}, \qquad g(x) = \frac{x^2}{2}.$$

Hence

$$\int_0^1 x\ln(x)\,dx = \frac{x^2}{2}\ln(x)\Big|_0^1 - \int_0^1 \frac{x^2}{2}\cdot\frac{1}{x}\,dx.$$

The boundary term is 0*, and the remaining integral is*

$$-\frac{1}{2}\int_0^1 x\,dx = -\frac{1}{2}\cdot\frac{1}{2} = -\frac{1}{4}.$$

Thus

$$\int_0^1 x\ln(x)\,dx = -\frac{1}{4}.$$

15.4.4 Partial Fraction Decomposition

For rational functions, that is, quotients of two polynomials,

$$r(x) = \frac{p(x)}{q(x)},$$

the direct search for an antiderivative often does not lead very far. However, once the denominator can be decomposed into simple factors, the fraction can be split into a sum of elementary fractions. Each of these partial fractions is then usually easy to integrate, for instance by means of logarithms or inverse trigonometric functions.

Thus, partial fraction decomposition is not a new integration theorem in the strict sense, but rather an algebraic tool that prepares the integration problem.

Theorem 523 (Partial fraction decomposition in the simple case). *Let $a_1, \ldots, a_n \in \mathbb{R}$ be pairwise distinct, and let*

$$q(x) = \prod_{k=1}^{n} (x - a_k).$$

If p is a polynomial with $\deg p < n$, then there exist unique numbers $c_1, \ldots, c_n \in \mathbb{R}$ such that

$$\frac{p(x)}{q(x)} = \sum_{k=1}^{n} \frac{c_k}{x - a_k}.$$

Proof. Multiplying the asserted equation by $q(x)$ gives

the polynomial identity

$$p(x) = \sum_{k=1}^{n} c_k \prod_{\substack{j=1 \\ j\neq k}}^{n} (x - a_j).$$

We now determine the coefficients c_k by substituting $x = a_k$. Then only the corresponding summand remains on the right-hand side, and hence

$$p(a_k) = c_k \prod_{\substack{j=1 \\ j\neq k}}^{n} (a_k - a_j).$$

Since the numbers $a_1, \ldots, a_n$ are pairwise distinct, this product is nonzero. Therefore

$$c_k = \frac{p(a_k)}{\prod_{j\neq k}(a_k - a_j)}.$$

Thus the coefficients are uniquely determined.

It remains to show that the entire identity follows from these coefficients. Consider the difference between the two sides after multiplication by $q(x)$. This gives a polynomial of degree at most $n-1$ that has the n pairwise distinct zeros $a_1, \ldots, a_n$. Such a polynomial can only be the zero polynomial. Hence the asserted decomposition indeed holds. □

Remark 524. *In general, if the denominator has multiple zeros, higher powers such as*

$$\frac{A}{x-a} + \frac{B}{(x-a)^2}$$

also occur. For the basic idea of the method, however, the simple case with pairwise distinct linear factors is sufficient at first.

Example 525. *We compute*

$$\int \frac{1}{x^2-1}\,dx.$$

Since

$$x^2-1=(x-1)(x+1),$$

we set

$$\frac{1}{x^2-1}=\frac{A}{x-1}+\frac{B}{x+1}.$$

After multiplication by $(x-1)(x+1)$, *we obtain*

$$1=A(x+1)+B(x-1).$$

Substituting $x=1$ *gives* $1=2A$, *hence* $A=\frac{1}{2}$. *Substituting* $x=-1$ *gives* $1=-2B$, *hence* $B=-\frac{1}{2}$. *Thus*

$$\frac{1}{x^2-1}=\frac{1}{2}\left(\frac{1}{x-1}-\frac{1}{x+1}\right).$$

It follows that

$$\begin{aligned}\int \frac{1}{x^2-1}\,dx &= \frac{1}{2}\int\left(\frac{1}{x-1}-\frac{1}{x+1}\right)dx\\ &= \frac{1}{2}\ln|x-1|-\frac{1}{2}\ln|x+1|+C\end{aligned}$$

Therefore

$$\int \frac{1}{x^2-1}\,dx=\frac{1}{2}\ln\left|\frac{x-1}{x+1}\right|+C.$$

Example 526. *The integral*

$$\int \frac{2x+3}{x^2+x-2}\,dx$$

can also be treated in this way. Since

$$x^2+x-2=(x+2)(x-1),$$

we look for a decomposition of the form

$$\frac{2x+3}{(x+2)(x-1)}=\frac{A}{x+2}+\frac{B}{x-1}.$$

After multiplying by $(x+2)(x-1)$, we obtain

$$2x+3=A(x-1)+B(x+2).$$

Substituting $x=1$ gives $5=3B$, hence $B=\frac{5}{3}$. Substituting $x=-2$ gives $-1=-3A$, hence $A=\frac{1}{3}$. It follows that

$$\int \frac{2x+3}{x^2+x-2}\,dx=\frac{1}{3}\int \frac{1}{x+2}\,dx+\frac{5}{3}\int \frac{1}{x-1}\,dx,$$

and therefore

$$\int \frac{2x+3}{x^2+x-2}\,dx=\frac{1}{3}\ln|x+2|+\frac{5}{3}\ln|x-1|+C.$$

15.4.5 Trigonometric Integrals

Trigonometric functions occur particularly often in integral calculus. Many integrals involving powers or products of sine and cosine can be greatly simplified by elementary transformations. The key idea is to rewrite trigonometric expressions into simpler terms by means of known identities and then apply the fundamental theorem of calculus or the substitution rule.

Proposition 527 (Auxiliary formulas for trigonometric integrals). *For all $x \in \mathbb{R}$, the following identities hold:*

$$\sin^2 x = \frac{1 - \cos(2x)}{2}$$

$$\cos^2 x = \frac{1 + \cos(2x)}{2}$$

$$\sin x \cos x = \frac{1}{2} \sin(2x)$$

Proof. From the addition formulas

$$\cos(x + y) = \cos x \cos y - \sin x \sin y$$

and

$$\sin(x + y) = \sin x \cos y + \cos x \sin y$$

we first obtain, by substituting $y = x$,

$$\cos(2x) = \cos^2 x - \sin^2 x$$

and

$$\sin(2x) = 2 \sin x \cos x.$$

Together with the well-known identity

$$\sin^2 x + \cos^2 x = 1$$

the three asserted formulas are obtained by simple addition and subtraction. □

These formulas are especially useful because they reduce powers of sine and cosine. For integrals of the form

$$\int \sin^m(x)\cos^n(x)\,dx$$

the following practical strategy is often useful:

- If one of the two exponents is odd, separate off one factor $\sin x$ or $\cos x$ and then use substitution.
- If both exponents are even, use the half-angle formulas from the proposition.

Example 528. *We compute*

$$\int \sin^3 x\cos^2 x\,dx.$$

Since the exponent of $\sin x$ *is odd, we write*

$$\sin^3 x = \sin x\cdot\sin^2 x = \sin x\,(1-\cos^2 x).$$

Thus

$$\int \sin^3 x\cos^2 x\,dx = \int \sin x\,(1-\cos^2 x)\cos^2 x\,dx.$$

Now set

$$u = \cos x, \qquad du = -\sin x\,dx.$$

It follows that

$$\int \sin^3 x \cos^2 x \, dx = -\int (1-u^2)u^2 \, du = -\int (u^2-u^4) \, du.$$

Thus we obtain

$$-\left(\frac{u^3}{3} - \frac{u^5}{5}\right) + C = -\frac{\cos^3 x}{3} + \frac{\cos^5 x}{5} + C.$$

Example 529. *For*

$$\int \sin^2 x \, dx$$

we use the half-angle identity

$$\sin^2 x = \frac{1 - \cos(2x)}{2}.$$

This gives

$$\int \sin^2 x \, dx = \frac{1}{2} \int (1 - \cos(2x)) \, dx = \frac{x}{2} - \frac{\sin(2x)}{4} + C.$$

Example 530. *The integral*

$$\int_0^\pi \sin x \cos x \, dx$$

can also be handled easily. Using

$$\sin x \cos x = \frac{1}{2} \sin(2x)$$

we obtain

$$\int_0^\pi \sin x \cos x \, dx = \frac{1}{2} \int_0^\pi \sin(2x) \, dx = -\frac{1}{4} \cos(2x) \Big|_0^\pi = 0.$$

Remark 531. *Trigonometric integrals are therefore usually not solved by a single general formula, but by suitable transformations. The key is to recognize the right identity or substitution.*

Remark 532. *Not all integrals can be expressed in terms of elementary antiderivatives. In such cases, numerical methods are often used to approximate integrals. Among the simplest methods are the rectangle rule and the trapezoidal rule.*

15.5 Exercises and Solutions

Exercise 533. *Use the Fundamental Theorem of Calculus to compute the definite integral*

$$\int_0^2 (3x^2 - 4x + 1)\, dx.$$

Solution 534. *An antiderivative of*

$$f(x) = 3x^2 - 4x + 1$$

is

$$F(x) = x^3 - 2x^2 + x.$$

Hence, by the Fundamental Theorem of Calculus,

$$\int_0^2 (3x^2 - 4x + 1)\, dx = F(2) - F(0).$$

Now

$$F(2) = 8 - 8 + 2 = 2, \qquad F(0) = 0.$$

Therefore,

$$\int_0^2 (3x^2 - 4x + 1)\, dx = 2.$$

Exercise 535. *Compute the integral by substitution:*

$$\int_0^1 2x\, e^{x^2}\, dx.$$

Solution 536. *We set*

$$u = x^2.$$

Then

$$du = 2x\, dx.$$

The limits also become

$$x = 0 \Rightarrow u = 0, \qquad x = 1 \Rightarrow u = 1.$$

Thus we obtain

$$\int_0^1 2x\, e^{x^2}\, dx = \int_0^1 e^u\, du.$$

An antiderivative of e^u *is* e^u*, so*

$$\int_0^1 e^u\, du = e - 1.$$

Therefore,

$$\int_0^1 2x\, e^{x^2}\, dx = e - 1.$$

Exercise 537. *Compute the integral using integration by parts:*

$$\int x \ln(x)\, dx.$$

Solution 538. *We choose*

$$f(x) = \ln(x), \qquad g'(x) = x.$$

Then

$$f'(x) = \frac{1}{x}, \qquad g(x) = \frac{x^2}{2}.$$

Using the formula for integration by parts, we obtain

$$\int x \ln(x)\, dx = \frac{x^2}{2} \ln(x) - \int \frac{x^2}{2} \cdot \frac{1}{x}\, dx.$$

Hence

$$\int x \ln(x)\, dx = \frac{x^2}{2} \ln(x) - \frac{1}{2} \int x\, dx = \frac{x^2}{2} \ln(x) - \frac{x^2}{4} + C.$$

Therefore,

$$\int x \ln(x)\, dx = \frac{x^2}{2} \ln(x) - \frac{x^2}{4} + C.$$

Exercise 539. *Compute the integral*

$$\int \frac{1}{x^2 - 1}\, dx$$

using a partial fraction decomposition.

Solution 540. *Since*

$$x^2 - 1 = (x-1)(x+1),$$

we set

$$\frac{1}{x^2-1} = \frac{A}{x-1} + \frac{B}{x+1}.$$

After multiplying by $(x-1)(x+1)$*, we obtain*

$$1 = A(x+1) + B(x-1).$$

Substituting $x = 1$ *gives*

$$1 = 2A \quad \Rightarrow \quad A = \frac{1}{2}.$$

Substituting $x = -1$ *gives*

$$1 = -2B \quad \Rightarrow \quad B = -\frac{1}{2}.$$

Thus

$$\frac{1}{x^2-1} = \frac{1}{2}\left(\frac{1}{x-1} - \frac{1}{x+1}\right).$$

It follows that

$$\int \frac{1}{x^2-1}\,dx = \frac{1}{2}\int \frac{1}{x-1}\,dx - \frac{1}{2}\int \frac{1}{x+1}\,dx.$$

Therefore, we obtain

$$\int \frac{1}{x^2-1}\,dx = \frac{1}{2}\ln|x-1| - \frac{1}{2}\ln|x+1| + C = \frac{1}{2}\ln\left|\frac{x-1}{x+1}\right| + C.$$

Exercise 541. *Compute*

$$\int_0^{\pi} \sin^2(x)\,dx.$$

Solution 542. *We use the half-angle identity*

$$\sin^2 x = \frac{1-\cos(2x)}{2}.$$

Thus the integral becomes

$$\int_0^{\pi} \sin^2(x)\,dx = \frac{1}{2}\int_0^{\pi} (1-\cos(2x))\,dx.$$

Now

$$\int_0^{\pi} 1\,dx = \pi$$

and

$$\int_0^{\pi} \cos(2x)\,dx = \frac{1}{2}\sin(2x)\bigg|_0^{\pi} = 0.$$

Therefore,

$$\int_0^{\pi} \sin^2(x)\,dx = \frac{1}{2}\pi = \frac{\pi}{2}.$$

Chapter 16

Improper Integrals

In the preceding chapters, we encountered the Riemann integral in its simplest and at the same time most important basic setting: the interval of integration was finite, and the function to be integrated was bounded and sufficiently regular on this interval. Under these assumptions, an integral can be described directly as a limit of Riemann sums or, equivalently, in terms of upper and lower sums.

In many analytic questions, however, this setting is not sufficient. Often one of two difficulties occurs: either the interval of integration is unbounded, or the integrand becomes unbounded at some point of the interval, respectively is not defined there. Such integrals can no longer be treated directly as ordinary Riemann integrals. Instead, they are understood as limits of Riemann integrals over suitable subintervals. These are called *improper integrals.*

Improper integrals are a central tool in analysis. They make it possible to treat integrals over unbounded intervals, integrals with singularities, and many limiting processes that arise in applications. At the same time, they are closely connected with the notions of limits already introduced: whether an improper integral has a well-defined value is determined by whether a certain limit exists.

16.1 Definition of improper integrals

In what follows, we distinguish three fundamental cases.

16.1.1 Infinite limits of integration

Definition 543. *Let $f : [a, \infty[\to \mathbb{R}$ be a function that is Riemann integrable on the interval $[a, R]$ for every $R > a$.*

If the limit

$$\lim_{R\to\infty} \int_a^R f(x)\,dx$$

exists and is finite, then the improper integral

$$\int_a^\infty f(x)\,dx$$

is called convergent. *In this case, one defines*

$$\int_a^\infty f(x)\,dx := \lim_{R\to\infty} \int_a^R f(x)\,dx.$$

If the limit does not exist or is infinite, then the integral is called divergent.

Analogously, for a function $f :]-\infty, a] \to \mathbb{R}$ that is Riemann integrable on every interval $[R, a]$ with $R < a$, one defines

$$\int_{-\infty}^{a} f(x)\,dx := \lim_{R \to -\infty} \int_{R}^{a} f(x)\,dx,$$

provided that this limit exists and is finite.

16.1.2 Undefined Integrands at an Endpoint of Integration

Definition 544. *Let $f :]a, b] \to \mathbb{R}$ be a function that is Riemann integrable on the interval $[a+\varepsilon, b]$ for every $0 < \varepsilon < b - a$.*

If the limit

$$\lim_{\varepsilon \downarrow 0} \int_{a+\varepsilon}^{b} f(x)\,dx$$

exists and is finite, then the improper integral

$$\int_{a}^{b} f(x)\,dx$$

is called convergent. *In this case one defines*

$$\int_{a}^{b} f(x)\,dx := \lim_{\varepsilon \downarrow 0} \int_{a+\varepsilon}^{b} f(x)\,dx.$$

If the limit does not exist or is infinite, then the integral is called divergent.

The idea is simple: one approaches the problematic point a from the right and considers the integrals over the truncated intervals $[a+\varepsilon, b]$.

Correspondingly, for a function $f : [a, b[\to \mathbb{R}$ that is Riemann integrable on the interval $[a, b-\varepsilon]$ for every $0 < \varepsilon < b-a$, the improper integral at the upper endpoint is defined by

$$\int_a^b f(x)\,dx := \lim_{\varepsilon\downarrow 0} \int_a^{b-\varepsilon} f(x)\,dx,$$

provided that this limit exists and is finite.

16.1.3 The General Case

Definition 545. *Let*

$$f :]a, b[\to \mathbb{R}, \qquad a \in \mathbb{R}\cup\{-\infty\}, \qquad b \in \mathbb{R}\cup\{\infty\},$$

be a function that is Riemann integrable on every compact subinterval

$$[\varphi, \psi] \subset]a, b[.$$

Furthermore, let $c \in]a, b[$ be chosen arbitrarily.

If the limits

$$\int_a^c f(x)\,dx := \lim_{\varphi\downarrow a} \int_\varphi^c f(x)\,dx$$

and

$$\int_c^b f(x)\,dx := \lim_{\psi\uparrow b} \int_c^\psi f(x)\,dx$$

exist and are both finite, then the improper integral

$$\int_a^b f(x)\,dx$$

is called convergent. *In this case one defines*

$$\int_a^b f(x)\,dx := \int_a^c f(x)\,dx + \int_c^b f(x)\,dx.$$

Otherwise, the integral is called divergent.

Remark 546. *If the improper integral is convergent, then its value does not depend on the choice of the intermediate point* $c \in]a, b[$.

This general case includes, in particular, integrals over intervals of the form $]a, \infty[$, $]-\infty, b[$, and $]-\infty, \infty[$, as well as integrals with singularities in the interior of the domain of integration.

16.2 First Examples

We now consider some simple examples that show how improper integrals can be computed as limits of ordinary integrals.

Example 547. *For* $p > 1$, *we have*

$$\int_1^\infty \frac{1}{x^p}\,dx = \lim_{R\to\infty} \int_1^R x^{-p}\,dx.$$

For $p \neq 1$, an antiderivative of x^{-p} is given by

$$\frac{x^{1-p}}{1-p}.$$

Therefore,

$$\int_1^\infty \frac{1}{x^p}\,dx = \lim_{R\to\infty}\left[\frac{x^{1-p}}{1-p}\right]_{x=1}^{x=R} = \lim_{R\to\infty}\frac{R^{1-p}-1}{1-p}.$$

Since for $p > 1$ the exponent $1-p$ is negative, we have $R^{1-p} \to 0$, and hence

$$\int_1^\infty \frac{1}{x^p}\,dx = \frac{1}{p-1}.$$

For $p \leq 1$, the integral diverges.

Example 548. *The integral*

$$\int_0^1 \frac{1}{\sqrt{x}}\,dx$$

is improper because the integrand is unbounded at $x = 0$. By definition, we write

$$\int_0^1 \frac{1}{\sqrt{x}}\,dx = \lim_{\varepsilon\downarrow 0}\int_\varepsilon^1 x^{-1/2}\,dx.$$

An antiderivative of $x^{-1/2}$ is $2\sqrt{x}$. Hence we obtain

$$\int_0^1 \frac{1}{\sqrt{x}}\,dx = \lim_{\varepsilon\downarrow 0}\left[2\sqrt{x}\right]_{x=\varepsilon}^{x=1} = \lim_{\varepsilon\downarrow 0} 2(1-\sqrt{\varepsilon}) = 2.$$

Thus the integral converges.

Example 549. *The integral*

$$\int_1^\infty \frac{1}{x}\, dx$$

is improper because of the infinite upper limit of integration. We have

$$\int_1^\infty \frac{1}{x}\, dx = \lim_{R\to\infty} \int_1^R \frac{1}{x}\, dx = \lim_{R\to\infty} \left[\ln x\right]_{x=1}^{x=R} = \lim_{R\to\infty} \ln R.$$

Since $\ln R \to \infty$ *as* $R \to \infty$*, the integral diverges.*

Example 550. *For*

$$\int_0^1 \frac{1}{x^\alpha}\, dx$$

with $\alpha \in \mathbb{R}$*, one obtains*

$$\int_0^1 x^{-\alpha}\, dx = \lim_{\varepsilon\downarrow 0} \int_\varepsilon^1 x^{-\alpha}\, dx.$$

For $\alpha \neq 1$*, an antiderivative is given by*

$$\frac{x^{1-\alpha}}{1-\alpha}$$

Hence it follows that

$$\int_0^1 \frac{1}{x^\alpha}\, dx = \lim_{\varepsilon\downarrow 0} \frac{1-\varepsilon^{1-\alpha}}{1-\alpha}.$$

The integral converges if and only if $\alpha < 1$*, and diverges for* $\alpha \geq 1$*.*

16.3 Outlook

Improper integrals are not only a natural endpoint of integral calculus, but also an important starting point for further investigations of convergence. In the following sections, we shall encounter criteria that often make it possible to decide whether such integrals converge or diverge. The same basic idea will appear again later in the context of series: comparison arguments are among the most important tools in analysis.

16.4 The Integral Test for Series

Improper integrals are not only an important topic in their own right; they also provide a very useful tool for studying infinite series. In particular, for suitable monotonically decreasing functions, one can compare the convergence of a series with the convergence of an improper integral. The following criterion is a classical and very effective example of this connection.

Theorem 551 (Integral test). *Let $f : [1, \infty[\to \mathbb{R}_+$ be a monotonically decreasing function. Then the series*

$$\sum_{n=1}^{\infty} f(n)$$

converges if and only if the improper integral

$$\int_1^{\infty} f(x)\,dx$$

converges.

Proof. Since f is monotonically decreasing and nonnegative, for every $n \in \mathbb{N}$ and every $x \in [n, n+1]$ we have

$$f(n+1) \leq f(x) \leq f(n).$$

Integrating this inequality over the interval $[n, n+1]$, we obtain

$$f(n+1) \leq \int_n^{n+1} f(x)\, dx \leq f(n).$$

Summing from $n = 1$ to N gives

$$\sum_{n=2}^{N+1} f(n) \leq \int_1^{N+1} f(x)\, dx \leq \sum_{n=1}^{N} f(n).$$

We now prove both directions.

1. Suppose that the improper integral converges. Then the sequence

$$I_N := \int_1^N f(x)\, dx$$

is monotonically increasing and, by assumption, bounded. Hence (I_N) converges. From the inequality above, for every N we obtain

$$\sum_{n=2}^{N} f(n) \leq \int_1^N f(x)\, dx \leq \int_1^\infty f(x)\, dx.$$

Thus the sequence of partial sums

$$S_N := \sum_{n=1}^{N} f(n)$$

is monotonically increasing and bounded. Consequently, the series $\sum_{n=1}^{\infty} f(n)$ converges.

2. Suppose that the series converges. Then the sequence of partial sums (S_N) is bounded. From

$$\int_1^N f(x)\,dx \leq \sum_{n=1}^{N-1} f(n)$$

it follows that the sequence (I_N) is bounded. Since (I_N) is monotonically increasing because $f \geq 0$, it has a finite limit. Therefore the improper integral converges.

This proves that the series converges if and only if the corresponding improper integral converges. □

Example 552. *The integral test immediately yields the classical result for the p-series*

$$\sum_{n=1}^{\infty} \frac{1}{n^p}.$$

For

$$f(x) = \frac{1}{x^p}$$

the function f is monotonically decreasing and nonnegative on $[1, \infty[$. Moreover,

$$\int_1^{\infty} \frac{1}{x^p}\,dx$$

converges if and only if $p > 1$. By the integral test, it follows that

$$\sum_{n=1}^{\infty} \frac{1}{n^p} \quad \text{converges if and only if} \quad p > 1.$$

In particular, the harmonic series

$$\sum_{n=1}^{\infty} \frac{1}{n},$$

diverges because the corresponding integral

$$\int_1^{\infty} \frac{1}{x}\, dx$$

diverges.

16.5 Exercises and Solutions

The following exercises are intended to apply the definitions of improper integrals in typical cases and to develop confidence in deciding questions of convergence.

Exercise 553. *For $p \in \mathbb{R}$, investigate the convergence of the improper integral*

$$\int_1^{\infty} \frac{1}{x^p}\, dx$$

and compute it in the case of convergence.

Solution 554. *For $p \neq 1$, an antiderivative of x^{-p} is given by*

$$\frac{x^{1-p}}{1-p}.$$

Hence

$$\int_1^{\infty} \frac{1}{x^p}\, dx = \lim_{R\to\infty} \int_1^{R} x^{-p}\, dx = \lim_{R\to\infty} \frac{R^{1-p} - 1}{1-p}.$$

If $p > 1$, then $1 - p < 0$, and therefore $R^{1-p} \to 0$. Thus the integral converges, and

$$\int_1^\infty \frac{1}{x^p}\,dx = \frac{1}{p-1}.$$

If $p \leq 1$, the integral diverges. For $p = 1$, one obtains

$$\int_1^R \frac{1}{x}\,dx = \ln R \to \infty,$$

and for $p < 1$, the term R^{1-p} grows without bound.

Exercise 555. *Investigate and compute the improper integral*

$$\int_0^1 \frac{1}{\sqrt{x}}\,dx.$$

Solution 556. *The integrand is unbounded at $x = 0$, so we write*

$$\int_0^1 \frac{1}{\sqrt{x}}\,dx = \lim_{\varepsilon \downarrow 0} \int_\varepsilon^1 x^{-1/2}\,dx.$$

An antiderivative of $x^{-1/2}$ is $2\sqrt{x}$. Hence

$$\int_0^1 \frac{1}{\sqrt{x}}\,dx = \lim_{\varepsilon \downarrow 0} \left[2\sqrt{x}\right]_\varepsilon^1 = \lim_{\varepsilon \downarrow 0} 2(1 - \sqrt{\varepsilon}) = 2.$$

Thus the integral converges.

Exercise 557. *Investigate and compute the improper integral*

$$\int_{-\infty}^\infty \frac{1}{1+x^2}\,dx.$$

Solution 558. *We choose the intermediate point* $c = 0$ *and write*

$$\int_{-\infty}^{\infty} \frac{1}{1+x^2}\,dx = \int_{-\infty}^{0} \frac{1}{1+x^2}\,dx + \int_{0}^{\infty} \frac{1}{1+x^2}\,dx.$$

An antiderivative of $\frac{1}{1+x^2}$ *is* $\arctan x$. *Therefore,*

$$\int_{0}^{\infty} \frac{1}{1+x^2}\,dx = \lim_{R\to\infty} \left[\arctan x\right]_0^R = \frac{\pi}{2},$$

and analogously,

$$\int_{-\infty}^{0} \frac{1}{1+x^2}\,dx = \lim_{R\to-\infty} \left[\arctan x\right]_R^0 = \frac{\pi}{2}.$$

Thus the integral converges, and it follows that

$$\int_{-\infty}^{\infty} \frac{1}{1+x^2}\,dx = \pi.$$

Chapter 17

Uniform Convergence

In the preceding chapters, we first considered limits of numerical sequences and later limits of functions. For a sequence of functions

$$(f_n)_{n\in\mathbb{N}}, \qquad f_n : D \to \mathbb{R},$$

with common domain D, it is natural first to examine convergence pointwise: for each fixed $x \in D$, we consider the numerical sequence

$$\big(f_n(x)\big)_{n\in\mathbb{N}}.$$

If this sequence converges for every $x \in D$ to a value $f(x)$, then we obtain a limiting function $f : D \to \mathbb{R}$. This is called *pointwise convergence.*

This idea is simple and natural, but for many questions in analysis it turns out to be too weak. Indeed, properties of functions are not automatically preserved under pointwise passage to the limit. For example, it may happen that all functions f_n are continuous, while the pointwise limiting function f is no longer continuous. Thus the essential issue is not only *that* the function values converge, but also *how* they approach the limiting value.

This is precisely where uniform convergence comes in. It requires that the functions f_n approach the limiting function not merely at each individual point, but in a uniform way on the entire domain. Intuitively, all function values should converge to f simultaneously and with the same prescribed accuracy.

This stronger form of convergence is especially important in analysis. Under suitable assumptions, continuity is preserved in the limit under uniform convergence, and it also plays a central role in integration and differentiation. It is therefore one of the most important tools for connecting the formation of limits with other operations.

17.1 Sequences of Functions

Definition 559 (Sequence of functions). *Let $D \subseteq \mathbb{R}$ be a set. A* sequence of functions *on D is a sequence*

$$(f_n)_{n\in\mathbb{N}}$$

of functions $f_n : D \to \mathbb{R}$.

A sequence of functions is therefore nothing other than a sequence of functions with a common domain. For each fixed $x \in D$, it gives rise to the numerical sequence

$$(f_n(x))_{n\in\mathbb{N}}.$$

Thus, questions about sequences of functions can often first be reduced to familiar statements about numerical sequences.

17.2 Pointwise Convergence

Definition 560 (Pointwise convergence). *Let $(f_n)_{n\in\mathbb{N}}$ be a sequence of functions on D, and let $f : D \to \mathbb{R}$ be a function. Then (f_n) is said to* converge pointwise *to f on D if, for every $x \in D$, the numerical sequence $(f_n(x))_{n\in\mathbb{N}}$ converges to $f(x)$. This means:*

$$\begin{aligned}
&\forall x \in D \quad \forall \varepsilon > 0 \\
&\exists N = N(x, \varepsilon) \in \mathbb{N} : \\
&\quad \forall n \geq N : \quad |f_n(x) - f(x)| < \varepsilon
\end{aligned}$$

In pointwise convergence, the required N may therefore depend on the point x under consideration. This is precisely where its weakness lies: at different points of the domain, the approximation to the limiting function may occur at very different speeds.

Remark 561. *Pointwise convergence therefore means: for each fixed $x \in D$, the sequence of function values $(f_n(x))$ converges to the value $f(x)$. The point x is held fixed, and only then is the limit process in n considered.*

17.3 Uniform Convergence

Definition 562 (Uniform convergence)**.** *Let* $(f_n)_{n\in\mathbb{N}}$ *be a sequence of functions on* D*, and let* $f : D \to \mathbb{R}$*. Then* (f_n) *is called* uniformly convergent *to* f *on* D *if*

$$\forall \varepsilon > 0\ \exists N \in \mathbb{N} :\ \forall n \geq N\ \forall x \in D :\ |f_n(x) - f(x)| < \varepsilon.$$

The difference from pointwise convergence lies solely in the order of the quantifiers: in uniform convergence, the number N may no longer depend on x. Thus a single N must suffice and work simultaneously for all points $x \in D$. Intuitively, the entire sequence of functions then converges *uniformly* to the limiting function on the whole domain.

Remark 563. *Uniform convergence is a substantially stronger requirement than pointwise convergence. It expresses that the functions* f_n *approximate the limiting function* f *not only at each individual point, but everywhere on* D *with the same accuracy.*

Remark 564. *Uniform convergence is often also written in the form*

$$\sup_{x\in D} |f_n(x) - f(x)| \to 0 \qquad (n \to \infty),$$

provided that the supremum is meaningfully defined. This notation makes the idea of uniform approximation particularly transparent.

17.4 First Guiding Idea

The main motivation for the notion of uniform convergence is the following: Pointwise convergence only says that, for each individual $x \in D$, the values $f_n(x)$ approach the limiting value $f(x)$. For many analytic properties, this is not sufficient. Uniform convergence, by contrast, requires a common control for all $x \in D$ at the same time. Precisely this stronger form of convergence is the right condition for preserving properties such as continuity under passage to the limit.

In the course of this chapter, we shall see that uniform convergence is of central importance in particular in three situations:

- when passing from continuous sequences of functions to their limit,
- when interchanging limits and integration,
- when interchanging limits and differentiation.

Thus uniform convergence becomes one of the most important notions in higher analysis.

17.5 Uniform Convergence and Continuity

Uniform convergence was introduced because pointwise convergence is too weak for many questions in analysis.

In particular, continuity can be lost when passing to the limit: even if all functions in a sequence are continuous, the pointwise limit function need not be continuous.

Uniform convergence remedies this problem. It ensures that the functions f_n approximate the limit function f simultaneously on the entire domain. This preserves enough control over the behavior of the functions to transfer continuity to the limit function.

The following theorem is therefore of fundamental importance. It shows that uniform convergence is precisely the right condition for preserving continuity under passage to the limit.

Theorem 565 (Preservation of continuity under uniform convergence). *Let $C \subset \mathbb{C}$, and let $(f_n)_{n\in\mathbb{N}}$ be a sequence of functions with*

$$f_n : C \to \mathbb{C}.$$

Assume that each function f_n is continuous and that f_n converges uniformly to a function

$$f : C \to \mathbb{C}.$$

Then the limit function f is continuous as well.

Proof. We prove the continuity of f at an arbitrary point $z_0 \in C$.

Let $\varepsilon > 0$ be given. Since (f_n) converges uniformly to f, there exists an $N \in \mathbb{N}$ such that

$$|f_N(z) - f(z)| < \frac{\varepsilon}{3} \qquad \text{for all } z \in C.$$

In particular, we also have

$$|f_N(z_0) - f(z_0)| < \frac{\varepsilon}{3}.$$

Since f_N is continuous at z_0, there exists a $\delta > 0$ such that

$$|f_N(z) - f_N(z_0)| < \frac{\varepsilon}{3} \qquad \text{for all } z \in C \text{ with } |z - z_0| < \delta.$$

For such z, the triangle inequality gives

$$\begin{aligned} |f(z) - f(z_0)| \le |f(z) - f_N(z)| \\ + |f_N(z) - f_N(z_0)| \\ + |f_N(z_0) - f(z_0)| \end{aligned}$$

Each of the three summands is smaller than $\varepsilon/3$. Hence

$$|f(z) - f(z_0)| < \varepsilon.$$

Thus we have shown that for every $\varepsilon > 0$ there exists a $\delta > 0$ such that

$$|z - z_0| < \delta \implies |f(z) - f(z_0)| < \varepsilon.$$

Therefore f is continuous at z_0. Since $z_0 \in C$ was arbitrary, f is continuous on C. □

Remark 566. *The proof illustrates the central idea of uniform convergence. First, one chooses one of the functions f_N that is already very close to the limit function f on the entire domain. Then one uses the* *continuity of f_N. Uniform convergence ensures that the deviations between f and f_N are small everywhere at the same time. This allows the continuity of f_N to be transferred to the limit function f.*

Without uniform convergence, this would not be possible in general, since the required approximation could vary in quality from point to point.

Example 567. *Consider the sequence of functions*

$$f_n : \mathbb{R} \to \mathbb{R}, \qquad f_n(x) = \frac{x}{1+nx^2}.$$

Each function f_n is continuous. Moreover,

$$|f_n(x)| = \frac{|x|}{1+nx^2}.$$

The maximum of this function is attained at $x = \frac{1}{\sqrt{n}}$ and is equal to

$$\frac{1}{2\sqrt{n}}.$$

Therefore,

$$\sup_{x\in\mathbb{R}} |f_n(x)| = \frac{1}{2\sqrt{n}} \to 0.$$

Thus (f_n) converges uniformly to the zero function. In particular, the limit function is continuous.

17.6 Supremum Norm and Uniform Convergence

Uniform convergence means that the deviation between f_n and the limiting function f becomes small simultaneously on the entire domain. This idea can be expressed especially clearly by considering not the deviation at a single point, but the *largest* deviation on the whole domain.

For this purpose, we introduce the supremum norm.

Definition 568 (Supremum norm). *Let $C \subset \mathbb{C}$, and let $\mathcal{B}(C)$ be the set of all bounded functions $f : C \to \mathbb{C}$. For $f \in \mathcal{B}(C)$, we define*

$$\|f\|_\infty := \sup_{z \in C} |f(z)|.$$

This number is called the supremum norm *of f.*

Remark 569. *The supremum norm measures the largest possible absolute value of a function. It is therefore a measure of how large a function can become on its entire domain.*

Remark 570. *The supremum norm is finite only for bounded functions. In the context of uniform convergence, however, this is not an obstacle, since under uniform convergence the differences $f_n - f$ are automatically bounded for sufficiently large n.*

With the help of the supremum norm, uniform convergence can be formulated very compactly.

Theorem 571 (Characterization of uniform convergence by the supremum norm). *Let $C \subset \mathbb{C}$, and let $f_n, f : C \to \mathbb{C}$ be functions. Then:*

$$f_n \to f \text{ uniformly on } C \quad \Longleftrightarrow \quad \|f_n - f\|_\infty \to 0.$$

Proof. We prove both directions.

„$\Rightarrow$" Assume that f_n converges uniformly to f on C. Then, by definition,

$$\forall \varepsilon > 0 \; \exists N \in \mathbb{N} : \; \forall n \geq N \; \forall z \in C : \; |f_n(z) - f(z)| < \varepsilon.$$

For fixed $n \geq N$, it follows that

$$\sup_{z \in C} |f_n(z) - f(z)| \leq \varepsilon,$$

hence

$$\|f_n - f\|_\infty \leq \varepsilon.$$

Since this holds for every $\varepsilon > 0$, we obtain $\|f_n - f\|_\infty \to 0$.

„$\Leftarrow$" Conversely, assume that $\|f_n - f\|_\infty \to 0$. Let $\varepsilon > 0$ be given. Then there exists an $N \in \mathbb{N}$ such that

$$\|f_n - f\|_\infty < \varepsilon \qquad \text{for all } n \geq N.$$

Then, for every $z \in C$,

$$|f_n(z) - f(z)| \leq \sup_{w \in C} |f_n(w) - f(w)| = \|f_n - f\|_\infty < \varepsilon.$$

Thus

$$\forall n \geq N \; \forall z \in C : \; |f_n(z) - f(z)| < \varepsilon,$$

and therefore f_n converges uniformly to f. □

Remark 572. *The supremum norm makes clear what uniform convergence means: what matters is not the* *error at a single point, but the largest error on the entire domain. This is precisely why the notation*

$$\|f_n - f\|_\infty \to 0$$

is so useful: it summarizes the entire definition in a single number.

Theorem 573 (Weierstrass convergence criterion). *Let $C \subset \mathbb{C}$, and let $(f_n)_{n\in\mathbb{N}}$ be a sequence of functions*

$$f_n : C \to \mathbb{C}.$$

Suppose that the series of supremum norms converges, that is,

$$\sum_{n=0}^{\infty} \|f_n\|_C < \infty.$$

Then the function series

$$\sum_{n=0}^{\infty} f_n$$

converges absolutely and uniformly on C to a function $F : C \to \mathbb{C}$.

Proof. For the partial sums, we set

$$S_N(z) := \sum_{n=0}^{N} f_n(z) \qquad (z \in C).$$

We have to show that (S_N) converges uniformly to a function F.

1. Pointwise absolute convergence. Let $z \in C$ be arbitrary. Then, for every $n \in \mathbb{N}$,

$$|f_n(z)| \leq \|f_n\|_C.$$

Since the real series $\sum_{n=0}^{\infty} \|f_n\|_C$ converges, the comparison test implies that

$$\sum_{n=0}^{\infty} |f_n(z)|$$

also converges. Hence the series $\sum_{n=0}^{\infty} f_n(z)$ converges absolutely for every $z \in C$. We may therefore define the limit function by

$$F(z) := \sum_{n=0}^{\infty} f_n(z)$$

.

2. Uniform convergence. Let $m > n$. Then, by the triangle inequality, for all $z \in C$,

$$\begin{aligned}|S_m(z) - S_n(z)| &= \left| \sum_{k=n+1}^{m} f_k(z) \right| \\ &\leq \sum_{k=n+1}^{m} |f_k(z)| \\ &\leq \sum_{k=n+1}^{m} \|f_k\|_C.\end{aligned}$$

Since the numerical series $\sum_{n=0}^{\infty} \|f_n\|_C$ converges, its tails become arbitrarily small. Thus

$$\sum_{k=n+1}^{m} \|f_k\|_C \to 0 \qquad \text{as } m, n \to \infty.$$

It follows that

$$\sup_{z\in C} |S_m(z) - S_n(z)| \leq \sum_{k=n+1}^{m} \|f_k\|_C \to 0.$$

Thus the partial sums (S_N) are uniformly Cauchy and hence converge uniformly to F.

Therefore the series $\sum_{n=0}^{\infty} f_n$ converges absolutely and uniformly to F. □

Remark 574. *The Weierstrass criterion is one of the most important tools for series of functions. It replaces the study of the series of functions by the study of the numerical series*

$$\sum_{n=0}^{\infty} \|f_n\|_C.$$

Since the norms $\|f_n\|_C$ are nonnegative real numbers, the convergence of the series of functions can be reduced to a question about ordinary numerical series. If the series of norms converges, then the absolute and uniform convergence of the series of functions follows automatically.

17.7 Uniform Convergence and Integration

One first important reason for introducing uniform convergence is that, under this assumption, the passage to the limit may be interchanged with integration. This is of great importance in analysis, since a function often does not appear directly, but rather as the limit function of a sequence of simpler functions. One then wants to know whether it is permissible to integrate first and only afterwards take the limit.

For pointwise convergence, this is generally not possible. For uniform convergence, however, the situation is much better: if a sequence of continuous functions on an interval converges uniformly, then the limit function is also continuous, and the integral may therefore be passed to the limit.

Theorem 575 (Interchange of integral and limit). *Let $(f_n)_{n\in\mathbb{N}}$ be a sequence of continuous functions*

$$f_n : [a,b] \to \mathbb{R},$$

which converges uniformly on $[a,b]$ to a function $f : [a,b] \to \mathbb{R}$. Then f is continuous, and

$$\int_a^b f(x)\,dx = \lim_{n\to\infty} \int_a^b f_n(x)\,dx.$$

Proof. By the theorem on the preservation of continuity under uniform convergence, the limit function f is continuous on $[a, b]$. Hence f is Riemann integrable on the interval $[a, b]$.

Since $f_n \to f$ uniformly on $[a, b]$, for every $\varepsilon > 0$ there exists an $N \in \mathbb{N}$ such that

$$|f_n(x) - f(x)| < \varepsilon \qquad \text{for all } x \in [a, b] \text{ and all } n \geq N.$$

In particular, the deviation between f_n and f is bounded by ε on the entire interval.

Now, for $n \geq N$, we consider the difference of the integrals. By the linearity of the integral, we obtain

$$\left| \int_a^b f_n(x)\, dx - \int_a^b f(x)\, dx \right| = \left| \int_a^b (f_n(x) - f(x))\, dx \right|.$$

Using the standard estimate for integrals, it follows that

$$\left| \int_a^b (f_n(x) - f(x))\, dx \right| \leq \int_a^b |f_n(x) - f(x)|\, dx.$$

Since the inequality $|f_n(x) - f(x)| < \varepsilon$ holds for all $x \in [a, b]$ and all $n \geq N$, we further obtain

$$\int_a^b |f_n(x) - f(x)|\, dx \leq \int_a^b \varepsilon\, dx = \varepsilon(b - a).$$

Thus, for all $n \geq N$,

$$\left| \int_a^b f_n(x)\, dx - \int_a^b f(x)\, dx \right| \leq \varepsilon(b - a).$$

Since $\varepsilon > 0$ was arbitrary, it follows that

$$\int_a^b f_n(x)\, dx \to \int_a^b f(x)\, dx.$$

This proves that

$$\int_a^b f(x)\,dx = \lim_{n\to\infty} \int_a^b f_n(x)\,dx.$$

□

Remark 576. *The idea of the proof is simple: because the functions f_n approximate the limit function f uniformly well on the whole interval, the error after integration also becomes uniformly small. The integral does not amplify this error in an uncontrolled way, but at most by the factor $b-a$.*

This is precisely the advantage of uniform convergence: it provides a global error control that is strong enough to interchange the passage to the limit with integration.

Example 577 (Pointwise convergence is generally not sufficient)**.** *We consider, on the interval $[0,1]$, the sequence of continuous functions*

$$f_n : [0,1] \to \mathbb{R},$$

defined by

$$f_n(x) = \begin{cases} n^2 x, & 0 \le x \le \frac{1}{n}, \\ n^2\left(\frac{2}{n} - x\right), & \frac{1}{n} \le x \le \frac{2}{n}, \\ 0, & \frac{2}{n} \le x \le 1. \end{cases}$$

Each function f_n is continuous.

For every fixed $x \in [0,1]$, *the following holds: if* n *is sufficiently large, then* $x > \frac{2}{n}$, *and hence* $f_n(x) = 0$. *Therefore* (f_n) *converges pointwise to the zero function*

$$f(x) = 0.$$

Thus

$$\int_0^1 f(x)\,dx = 0.$$

On the other hand, the graph of f_n *is a triangle with base length* $\frac{2}{n}$ *and height* n. *Hence*

$$\int_0^1 f_n(x)\,dx = \frac{1}{2} \cdot \frac{2}{n} \cdot n = 1 \qquad \textit{for all } n \in \mathbb{N}.$$

It follows that

$$\lim_{n\to\infty} \int_0^1 f_n(x)\,dx = 1 \neq 0 = \int_0^1 f(x)\,dx.$$

This shows that, under pointwise convergence, one may generally not interchange the limit with the integral.

Example 578 (Interchanging the integral and the limit). *We consider, on the interval* $[0,1]$, *the sequence of functions*

$$f_n : [0,1] \to \mathbb{R}, \qquad f_n(x) = \frac{x}{n}.$$

For every $x \in [0,1]$, *we have*

$$|f_n(x)| = \frac{x}{n} \leq \frac{1}{n}.$$

Thus

$$\sup_{x\in[0,1]} |f_n(x)| \leq \frac{1}{n} \to 0.$$

Consequently, (f_n) *converges uniformly to the zero function*

$$f(x) = 0.$$

Now we compute the integrals:

$$\int_0^1 f_n(x)\,dx = \int_0^1 \frac{x}{n}\,dx = \frac{1}{n}\int_0^1 x\,dx = \frac{1}{n}\cdot\frac{1}{2} = \frac{1}{2n}.$$

Therefore

$$\lim_{n\to\infty}\int_0^1 f_n(x)\,dx = \lim_{n\to\infty}\frac{1}{2n} = 0.$$

On the other hand,

$$\int_0^1 f(x)\,dx = \int_0^1 0\,dx = 0.$$

Thus we obtain

$$\int_0^1 f(x)\,dx = \lim_{n\to\infty}\int_0^1 f_n(x)\,dx.$$

This example shows that, under uniform convergence, the passage to the limit may be interchanged with the integral.

17.8 Uniform Convergence and Differentiation

For differentiation, the situation is more delicate than for integration. In the case of integrals, uniform convergence of the sequence of functions itself is sufficient to interchange taking limits and integration. For derivatives, however, this is not enough. A sequence of continuous functions may even converge uniformly to a limiting function without that limiting function being differentiable.

A simple example is the sequence

$$f_n(x) = \sqrt{x^2 + \frac{1}{n}} \qquad (x \in [-1, 1]).$$

Then f_n converges uniformly to the function $f(x) = |x|$, but this function is not differentiable at $x = 0$. Thus uniform convergence of the functions themselves is not sufficient for differentiation.

In order to interchange the limiting process with differentiation, more control is needed, namely control over the derivatives as well.

Theorem 579 (Interchange of Limit and Derivative). *Let $(f_n)_{n\in\mathbb{N}}$ be a sequence of continuously differentiable functions*

$$f_n : [a, b] \to \mathbb{R},$$

and let $f : [a, b] \to \mathbb{R}$ be the pointwise limiting function of (f_n). Assume that the sequence of derivatives (f_n') converges uniformly on $[a, b]$ to a function

$$g : [a, b] \to \mathbb{R}.$$

Then f is continuously differentiable, and

$$f'(x) = g(x) = \lim_{n\to\infty} f_n'(x) \qquad \text{for all } x \in [a,b].$$

Proof. Since f_n' is continuous and (f_n') converges uniformly to g, the theorem on the preservation of continuity under uniform convergence implies that g is continuous as well.

Now fix some $x \in [a,b]$. For every $n \in \mathbb{N}$, the fundamental theorem of calculus gives

$$f_n(x) = f_n(a) + \int_a^x f_n'(t)\,dt.$$

Since f_n converges pointwise to f, in particular

$$f_n(a) \to f(a).$$

Moreover, (f_n') converges uniformly to g. Hence, by the previously proved theorem on interchanging limits and integrals,

$$\int_a^x f_n'(t)\,dt \to \int_a^x g(t)\,dt.$$

Thus we may pass to the limit $n \to \infty$ in the identity for $f_n(x)$ and obtain

$$f(x) = f(a) + \int_a^x g(t)\,dt.$$

Thus f is represented as the integral function of a continuous function g. By the fundamental theorem of calculus, f is therefore differentiable, and

$$f'(x) = g(x) \qquad \text{for all } x \in [a,b].$$

Since $g(x) = \lim_{n\to\infty} f_n'(x)$, it follows altogether that

$$f'(x) = \lim_{n\to\infty} f_n'(x) \qquad \text{for all } x \in [a,b].$$

□

Remark 580. *The proof makes the basic idea very clear: the uniform convergence of the derivatives makes it possible to control the derivatives under the integral. In this way, the limiting function is recognized as the integral function of a continuous function, and its differentiability follows.*

For differentiation, the convergence of the functions themselves is therefore not sufficient. What is also essential is that the derivatives converge in a controlled way.

Example 581 (Interchanging limits and differentiation). *We consider on the interval $[0,1]$ the sequence*

$$f_n(x) := \sum_{k=0}^{n} \frac{x^k}{k!}.$$

These are the partial sums of the exponential series.

First, (f_n) converges uniformly to the function

$$f(x) = e^x.$$

Indeed, for $x \in [0,1]$ we have

$$|e^x - f_n(x)| = \left| \sum_{k=n+1}^{\infty} \frac{x^k}{k!} \right| \le \sum_{k=n+1}^{\infty} \frac{1}{k!},$$

and the right-hand side tends to 0 as $n \to \infty$.

We now compute the derivatives:

$$f_n'(x) = \sum_{k=1}^{n} \frac{kx^{k-1}}{k!} = \sum_{k=1}^{n} \frac{x^{k-1}}{(k-1)!} = \sum_{k=0}^{n-1} \frac{x^k}{k!}.$$

Hence

$$f_n'(x) \to e^x \qquad \textit{uniformly on } [0,1],$$

since, again, for $x \in [0,1]$ we have

$$|e^x - f_n'(x)| = \left| \sum_{k=n}^{\infty} \frac{x^k}{k!} \right| \le \sum_{k=n}^{\infty} \frac{1}{k!} \to 0.$$

Thus the assumptions of the preceding theorem are satisfied. Consequently, the limiting function $f(x) = e^x$ is differentiable, and

$$f'(x) = \lim_{n\to\infty} f_n'(x) = e^x \qquad \textit{for all } x \in [0,1].$$

17.9 Power Series

After seeing in the previous section that uniform convergence behaves well with integration and differentiation, we now apply this theory to a particularly important class of functions: *power series.* They play a central role in analysis, because they allow complicated functions to be represented by series and reduce many operations to simple algebraic manipulations.

The geometric series

$$\sum_{n=0}^{\infty} x^n$$

is already a first example of a power series. It shows that infinite sums of powers are not only formally interesting, but can also define very concrete functions. In general, one considers series of the form

$$\sum_{n=0}^{\infty} a_n(x - x_0)^n,$$

where $(a_n)_{n\in\mathbb{N}_0}$ is a fixed sequence of real numbers and $x_0 \in \mathbb{R}$ is a prescribed point.

17.9.1 Definition and First Examples

Definition 582 (Power series). *A series of the form*

$$\sum_{n=0}^{\infty} a_n(x - x_0)^n$$

is called a power series centered at x_0. *The numbers* a_n *are called the* coefficients *of the power series.*

The term "power series" emphasizes that each summand contains only a power of $(x - x_0)$. In this sense, a power series is an infinite generalization of a polynomial. Whereas a polynomial has only finitely many terms, a power series may contain infinitely many terms.

The most important question is now:

For which values of x does the power series converge?

The first decisive observation is that this question can be answered not only pointwise, but structurally.

17.9.2 A first convergence theorem

We begin with a simple but fundamental theorem. It shows that convergence of a power series at a single point already implies convergence throughout an entire interval.

Theorem 583. *If the power series*

$$\sum_{n=0}^{\infty} a_n(x - x_0)^n$$

converges at a point $x_1 \neq x_0$*, then it even converges absolutely for all* x *satisfying*

$$|x - x_0| < |x_1 - x_0|.$$

Proof. Since the series converges at the point x_1, we have in particular

$$a_n(x_1 - x_0)^n \longrightarrow 0 \qquad (n \to \infty).$$

Thus the sequence $(a_n(x_1 - x_0)^n)$ is bounded. Hence there exists a constant $M > 0$ such that

$$|a_n(x_1 - x_0)^n| \leq M \qquad \text{for all } n \in \mathbb{N}_0.$$

Now let x be a point with

$$|x - x_0| < |x_1 - x_0|.$$

Then

$$q := \frac{|x - x_0|}{|x_1 - x_0|} \in [0, 1).$$

For every $n \in \mathbb{N}_0$, we obtain

$$|a_n(x - x_0)^n| = |a_n(x_1 - x_0)^n| \cdot \left(\frac{|x - x_0|}{|x_1 - x_0|}\right)^n \leq Mq^n.$$

Thus the series is absolutely dominated by the geometric series

$$\sum_{n=0}^{\infty} Mq^n,$$

which converges because $0 \leq q < 1$. Therefore

$$\sum_{n=0}^{\infty} |a_n(x - x_0)^n|$$

also converges, and hence the power series itself converges absolutely. □

This theorem is already very informative: convergence at one point implies convergence at all points with smaller distance from the center of expansion. Thus the set of convergence points cannot have an arbitrary shape; it has a clear geometric structure.

17.9.3 The Radius of Convergence

It follows immediately from the preceding theorem that every power series has a boundary for its region of convergence. This boundary is described by a radius.

Theorem 584 (Existence of a radius of convergence). *For every power series*

$$\sum_{n=0}^{\infty} a_n (x - x_0)^n$$

there exists a number

$$R \in [0, \infty]$$

with the following properties:

$$\textit{if } |x - x_0| < R, \ \textit{the series converges absolutely,}$$

$$\textit{if } |x - x_0| > R, \ \textit{the series diverges.}$$

Proof. We consider the set

$$K := \{\, |x - x_0| \ : \ \sum_{n=0}^{\infty} a_n (x - x_0)^n \text{ converges} \,\} \subseteq [0, \infty].$$

If the power series converges only for $x = x_0$, then $K = \{0\}$, and we set $R = 0$. Otherwise, K contains a positive number. By the preceding theorem, the following then holds: if a distance $r \in K$ occurs, then all smaller distances also belong to K. Thus the set of convergence distances is downward closed.

We now define

$$R := \sup K \in [0, \infty].$$

If $|x - x_0| < R$, then, by the definition of the supremum, there exists an $r \in K$ such that

$$|x - x_0| < r.$$

Since the series converges at a point with distance r, it also converges at x by the first theorem, and in fact absolutely.

Conversely, if $|x - x_0| > R$, then the series cannot converge at x, because otherwise $|x - x_0| \in K$ and hence $|x - x_0| \leq R$, contradicting the assumption.

This proves both assertions. □

The number R is called the *radius of convergence* of the power series. It describes the region in which the power series certainly converges and, at the same time, the region in which it certainly diverges.

It is important to note that this theorem says nothing yet about the boundary points

$$|x - x_0| = R$$

At such points, convergence may occur, but divergence may occur as well; both cases are possible. The precise investigation of the boundary is a separate question and will be treated later.

17.9.4 First Orientation

For the further development of the theory, the following idea is now essential:

> Inside the radius of convergence, power series behave particularly well.

It is precisely there that we shall later apply the theorems on uniform convergence and show that power series are very regular within their domain of convergence: they are continuous, and one may integrate and differentiate them term by term. In this way, the power series becomes an extremely powerful tool in analysis.

17.9.5 Uniform Convergence Inside the Radius of Convergence

The essential property of power series is that they behave particularly well *inside their radius of convergence.* More precisely: as soon as one stays a small distance away from the boundary of the domain of convergence, one obtains not only pointwise convergence, but even uniform convergence on every closed subinterval.

This is crucial for the further theory, since it is exactly here that we may apply the previously proved theorems on uniform convergence.

Theorem 585 (Uniform convergence of power series). *Let*

$$f(x) = \sum_{n=0}^{\infty} a_n (x - x_0)^n$$

be a power series with radius of convergence $R > 0$. Then, for every

$$0 < r < R$$

the series converges absolutely and uniformly on the closed interval

$$I_r := [x_0 - r, x_0 + r]$$

Proof. Let $0 < r < R$ be fixed. Then the numerical series

$$\sum_{n=0}^{\infty} |a_n| \, r^n$$

converges by the definition of the radius of convergence.

Now, for every $x \in I_r$, we have

$$|a_n(x - x_0)^n| \leq |a_n| \, r^n.$$

Thus the series of functions is majorized by the convergent numerical series

$$\sum_{n=0}^{\infty} |a_n| \, r^n$$

By the Weierstrass majorant test, the power series therefore converges uniformly on I_r.

Since the majorant series even controls the convergence of the absolute values, the convergence on I_r is also absolute. □

This theorem already illustrates an important basic fact: inside the radius of convergence, a power series is not merely convergent; it behaves so regularly that it can be treated like a very well-controlled sequence of polynomials.

Remark 586. *The endpoint case $|x - x_0| = R$ is deliberately not considered here. At the boundary of the domain of convergence, many different things may happen: the series may converge, diverge, or converge only conditionally.*

17.9.6 Termwise Differentiation of Power Series

We now come to perhaps the most important property of power series: within their radius of convergence, they may be differentiated term by term. This is the point at which the previously developed theory of uniform convergence becomes particularly effective.

Theorem 587 (Derivative of a power series). *Let*

$$f(x) = \sum_{n=0}^{\infty} a_n (x - x_0)^n$$

be a power series with radius of convergence $R > 0$. Then the formally differentiated series

$$g(x) := \sum_{n=1}^{\infty} n a_n (x - x_0)^{n-1}$$

also has radius of convergence R.

Moreover, g converges absolutely and uniformly on every interval

$$I_r = [x_0 - r, x_0 + r] \qquad (0 < r < R)$$

and the function defined by f is differentiable there, with

$$f'(x) = \sum_{n=1}^{\infty} n a_n (x - x_0)^{n-1}.$$

Proof. We first consider a fixed r with $0 < r < R$.

Since the series

$$\sum_{n=0}^{\infty} |a_n| \, r^n$$

converges, its terms are in particular bounded. Hence there exists a constant $M > 0$ such that

$$|a_n| \, r^n \leq M \qquad \text{for all } n \in \mathbb{N}_0.$$

For $x \in I_r$ we then have

$$|n a_n (x - x_0)^{n-1}| \leq n |a_n| \, r^{n-1}.$$

To control the right-hand side, choose a number ρ with

$$r < \rho < R.$$

Then

$$|a_n| \, \rho^n$$

is also bounded by a constant, and since

$$\frac{r}{\rho} < 1$$

the differentiated geometric series

$$\sum_{n=1}^{\infty} n \left(\frac{r}{\rho} \right)^{n-1}$$

converges. Therefore

$$\sum_{n=1}^{\infty} n|a_n|\, r^{n-1}$$

also converges. By the Weierstrass majorant criterion, it follows that g converges absolutely and uniformly on I_r.

Now set

$$f_N(x) := \sum_{n=0}^{N} a_n(x - x_0)^n.$$

Then each f_N is a polynomial, hence differentiable, and

$$f_N'(x) = \sum_{n=1}^{N} na_n(x - x_0)^{n-1}.$$

The sequence $(f_N')_{N\in\mathbb{N}}$ converges uniformly to g on I_r, and the sequence $(f_N(x_0))_{N\in\mathbb{N}}$ clearly converges to $f(x_0)$.

By the theorem on termwise differentiation of uniformly convergent sequences of functions, it follows that f is differentiable on I_r and that

$$f'(x) = g(x) = \sum_{n=1}^{\infty} na_n(x - x_0)^{n-1}.$$

It remains to show that g has the same radius of convergence as f.

From the convergence theorem for power series already proved, we first obtain the following: if g converges at a point $x_1 \neq x_0$, then g converges absolutely for all x satisfying

$$|x - x_0| < |x_1 - x_0|.$$

In particular, g therefore converges absolutely and uniformly on every smaller closed interval. By integrating term by term, one then obtains the original series

$$\sum_{n=0}^{\infty} a_n (x - x_0)^n$$

again as an antiderivative series of g. Consequently, the radius of convergence of g cannot be larger than that of f.

On the other hand, we have already shown above that g converges absolutely and uniformly on every interval I_r with $r < R$. Thus g has radius of convergence at least R.

Together, this shows that g has exactly the same radius of convergence R. □

Remark 588. *Thus, inside its radius of convergence, a power series is not merely a convergent series, but a very stable representation: there one may differentiate it term by term, and the differentiated series is again a power series with the same radius of convergence.*

17.9.7 Consequence: Term-by-Term Integration

For the same reason, power series may also be integrated term by term inside their radius of convergence. If

$$f(x) = \sum_{n=0}^{\infty} a_n(x - x_0)^n$$

and if $0 < r < R$, then on I_r one has

$$\int f(x)\,dx = C + \sum_{n=0}^{\infty} \frac{a_n}{n+1}(x - x_0)^{n+1}.$$

This too is an immediate consequence of uniform convergence.

This reveals the central idea behind power series: inside their radius of convergence, power series behave so regularly that one can calculate with them as if they were polynomials.

17.9.8 Examples

The general theory of power series becomes truly transparent only when it is tested on concrete series. The classical examples in particular show that the theorems just proved are not merely formal statements, but provide genuine computational tools.

Example 589 (The geometric series). *We first consider the power series*

$$f(x) = \sum_{n=0}^{\infty} x^n.$$

Here the center of expansion is $x_0 = 0$*, and* $a_n = 1$ *for all* n*.*

For $|x| < 1$*, we already know the convergence of the geometric series:*

$$\sum_{n=0}^{\infty} x^n = \frac{1}{1-x}.$$

Thus the radius of convergence is

$$R = 1.$$

We now apply the general theorems. Let $0 < r < 1$*. Then, for all* $x \in [-r, r]$*, we have*

$$|x|^n \leq r^n.$$

Since the numerical series

$$\sum_{n=0}^{\infty} r^n$$

converges, the Weierstrass majorant criterion implies that

$$\sum_{n=0}^{\infty} x^n$$

converges absolutely and uniformly on $[-r, r]$*.*

Termwise integration gives

$$\int_0^x \sum_{n=0}^{\infty} t^n \, dt = \sum_{n=0}^{\infty} \int_0^x t^n \, dt = \sum_{n=0}^{\infty} \frac{x^{n+1}}{n+1}.$$

On the other hand,

$$\int_0^x \frac{1}{1-t}\,dt = -\ln(1-x).$$

Thus, for $|x| < 1$,

$$-\ln(1-x) = \sum_{n=1}^{\infty} \frac{x^n}{n}.$$

Thus, from a single example, we have already obtained two important series: the geometric series itself and the logarithmic series.

Example 590 (The exponential series)**.** *We now consider the power series*

$$f(x) = \sum_{n=0}^{\infty} \frac{x^n}{n!}.$$

We first determine the radius of convergence. For the coefficients we have

$$\frac{a_{n+1}}{a_n} = \frac{1}{n+1}.$$

Hence, for the terms,

$$\left| \frac{x^{n+1}/(n+1)!}{x^n/n!} \right| = \frac{|x|}{n+1} \xrightarrow[n\to\infty]{} 0.$$

Thus the series converges for every $x \in \mathbb{R}$, *and its radius of convergence is*

$$R = \infty.$$

Therefore, we may apply the general theorems on every compact interval. In particular, the series converges absolutely and uniformly on every interval $[-r, r]$.

We now differentiate term by term:

$$\left(\sum_{n=0}^{\infty} \frac{x^n}{n!}\right)' = \sum_{n=1}^{\infty} \frac{nx^{n-1}}{n!} = \sum_{n=1}^{\infty} \frac{x^{n-1}}{(n-1)!}.$$

With the index shift $m = n - 1$, *we obtain*

$$\sum_{n=1}^{\infty} \frac{x^{n-1}}{(n-1)!} = \sum_{m=0}^{\infty} \frac{x^m}{m!}.$$

Therefore,

$$f'(x) = f(x).$$

Moreover,

$$f(0) = \sum_{n=0}^{\infty} \frac{0^n}{n!} = 1.$$

Thus the exponential series defines a function that is equal to its own derivative and takes the value 1 *at* 0. *This is a particularly striking example of how power series naturally solve differential equations.*

Example 591 (The logarithmic series)**.** *We return to the geometric series and integrate it term by term on* $[-r, r]$ *with* $0 < r < 1$*:*

$$\int_0^x \sum_{n=0}^{\infty} t^n \, dt = \sum_{n=0}^{\infty} \int_0^x t^n \, dt = \sum_{n=0}^{\infty} \frac{x^{n+1}}{n+1}.$$

Since the left-hand side is

$$\int_0^x \frac{1}{1-t}\,dt = -\ln(1-x)$$

we obtain

$$-\ln(1-x) = \sum_{n=1}^{\infty} \frac{x^n}{n} \qquad \text{for } |x| < 1.$$

Thus this series has the same radius of convergence as the geometric series, namely

$$R = 1.$$

This is also a good example of the fact that term-by-term integration does not change the radius of convergence.

For the derivative, we obtain directly

$$\frac{d}{dx}(-\ln(1-x)) = \frac{1}{1-x},$$

and on the right-hand side

$$\left(\sum_{n=1}^{\infty} \frac{x^n}{n}\right)' = \sum_{n=1}^{\infty} x^{n-1} = \sum_{n=0}^{\infty} x^n.$$

This agrees exactly with the geometric series.

Example 592 (The arctangent series)**.** *Here, too, we begin with a familiar geometric idea: For $|x| < 1$, we have*

$$\frac{1}{1+x^2} = \sum_{n=0}^{\infty} (-1)^n x^{2n}.$$

Indeed, this is simply the geometric series with $-x^2$ *in place of* x.

Since, for $0 < r < 1$,

$$|(-1)^n x^{2n}| \le r^{2n} \qquad (x \in [-r, r]),$$

the series converges absolutely and uniformly on $[-r, r]$.

Now we integrate term by term:

$$\int_0^x \frac{1}{1+t^2}\, dt = \sum_{n=0}^{\infty} (-1)^n \int_0^x t^{2n}\, dt = \sum_{n=0}^{\infty} \frac{(-1)^n}{2n+1} x^{2n+1}.$$

By definition, the left-hand side is

$$\arctan(x),$$

and therefore we obtain

$$\arctan(x) = \sum_{n=0}^{\infty} \frac{(-1)^n}{2n+1} x^{2n+1} \qquad \textit{for } |x| < 1.$$

Here again, the radius of convergence is equal to 1.

This example is particularly instructive because it shows how a transcendental function arises from a simple rational function by means of a series expansion.

Example 593 (The binomial series). *Finally, we consider a series that extends the notion of binomial coefficients to non-integer exponents. For a real exponent* α, *we define*

$$\binom{\alpha}{n} = \frac{\alpha(\alpha-1)\cdots(\alpha-n+1)}{n!} \qquad (n \in \mathbb{N},\ n \ge 1),$$

and

$$\binom{\alpha}{0} = 1.$$

The binomial series is then given by

$$(1+x)^\alpha = \sum_{n=0}^{\infty} \binom{\alpha}{n} x^n \qquad \text{for } |x| < 1.$$

Why is the radius of convergence equal to 1*? To see this, consider the ratio test:*

$$\frac{\binom{\alpha}{n+1} x^{n+1}}{\binom{\alpha}{n} x^n} = \frac{\alpha - n}{n+1} x.$$

For large n*, this quotient behaves like*

$$-x,$$

so its absolute value is asymptotically about $|x|$*. Hence the series converges for* $|x| < 1$ *and diverges for* $|x| > 1$.

Therefore, on every interval $[-r, r]$ *with* $0 < r < 1$*, the series converges absolutely and uniformly. In particular, it may be differentiated and integrated term by term there.*

This example illustrates particularly well how far the principle of power series reaches: even powers with arbitrary real exponents can be represented by series inside $(-1, 1)$.

Remark 594. *The examples reveal a common pattern: First one determines the radius of convergence, then one* *uses uniform convergence on smaller intervals, and finally one obtains new series and new function representations by integrating or differentiating term by term. This is precisely where the strength of power series methods lies.*

17.10 Exercises with Solutions

Exercise 595. *Investigate the sequence of functions*

$$f_n : [0,1] \to \mathbb{R}, \qquad f_n(x) = x^n,$$

with respect to pointwise convergence. Does (f_n) also converge uniformly on $[0,1]$?

Solution 596. *For every fixed $x \in [0,1)$, we have $x^n \to 0$ as $n \to \infty$, whereas for $x = 1$ we always have*

$$f_n(1) = 1$$

Thus the pointwise limit function is

$$f(x) = \begin{cases} 0, & 0 \leq x < 1, \\ 1, & x = 1. \end{cases}$$

This limit function is not continuous at $x = 1$. Since the functions f_n are continuous, the theorem on the preservation of continuity under uniform convergence implies that uniform convergence on $[0,1]$ cannot occur.

This can also be seen directly: For every n and every $x \in [0,1)$, we have

$$|f_n(x) - f(x)| = x^n.$$

As x approaches 1, the value x^n becomes arbitrarily close to 1. Hence

$$\sup_{x\in[0,1]} |f_n(x) - f(x)| = 1$$

for all n. Therefore, (f_n) does not converge uniformly on $[0,1]$.

Exercise 597. *Show that the sequence*

$$f_n : [0,r] \to \mathbb{R}, \qquad f_n(x) = x^n,$$

converges uniformly to 0 for every fixed $r \in (0,1)$. Also determine $\|f_n\|_\infty$ on $[0,r]$.

Solution 598. *For $x \in [0,r]$, we have*

$$0 \le x^n \le r^n.$$

It follows that

$$\sup_{x\in[0,r]} |f_n(x) - 0| = \sup_{x\in[0,r]} x^n = r^n.$$

Since $0 < r < 1$, we have $r^n \to 0$. Thus f_n converges uniformly to the zero function on $[0,r]$.

Moreover,

$$\|f_n\|_\infty = r^n.$$

Exercise 599. *Investigate the sequence*

$$f_n : [0,1] \to \mathbb{R}, \qquad f_n(x) = \frac{x}{n+x},$$

with respect to uniform convergence. Determine the limit and the supremum norm of f_n on $[0,1]$.

Solution 600. *For every fixed* $x \in [0,1]$*, we have*

$$\frac{x}{n+x} \to 0 \qquad (n \to \infty).$$

Thus the pointwise limit function is $f(x) = 0$.

To study uniform convergence, we consider

$$|f_n(x) - 0| = \frac{x}{n+x}.$$

The function $x \mapsto \frac{x}{n+x}$ *is monotonically increasing on* $[0,1]$*, so its maximum is attained at* $x = 1$*. Hence*

$$\|f_n\|_\infty = \sup_{x\in[0,1]} \frac{x}{n+x} = \frac{1}{n+1}.$$

Since $\frac{1}{n+1} \to 0$*, it follows that*

$$\|f_n - f\|_\infty \to 0.$$

Thus (f_n) *converges uniformly to* 0 *on* $[0,1]$.

Exercise 601. *Let*

$$f_n(x) = \frac{\sin x}{n} \qquad (x \in \mathbb{R}).$$

Show that (f_n) *converges uniformly to* 0.

Solution 602. *For all* $x \in \mathbb{R}$*, we have*

$$|f_n(x)| = \frac{|\sin x|}{n} \leq \frac{1}{n},$$

since $|\sin x| \leq 1$. *Hence*

$$\sup_{x \in \mathbb{R}} |f_n(x)| \leq \frac{1}{n}.$$

Since $\frac{1}{n} \to 0$, *it follows that*

$$\|f_n\|_\infty \to 0.$$

Thus (f_n) *converges uniformly to the zero function on* $\mathbb{R}$.

Exercise 603. *Show that the sequence*

$$f_n(x) = \sqrt{x^2 + \frac{1}{n}} \qquad (x \in [-1, 1])$$

converges uniformly to $|x|$. *Why does this example also show that uniform convergence need not preserve differentiability in the limit?*

Solution 604. *For every* $x \in [-1, 1]$, *we have*

$$f_n(x) = \sqrt{x^2 + \frac{1}{n}} \to |x|.$$

Thus the pointwise limit function is $f(x) = |x|$.

To prove uniform convergence, we consider the difference

$$\sqrt{x^2 + \frac{1}{n}} - |x| = \frac{1/n}{\sqrt{x^2 + \frac{1}{n}} + |x|}.$$

The denominator is at least $\sqrt{1/n}$, *and therefore*

$$0 \leq \sqrt{x^2 + \frac{1}{n}} - |x| \leq \frac{1/n}{\sqrt{1/n}} = \frac{1}{\sqrt{n}}.$$

Hence

$$\sup_{x\in[-1,1]} \big|f_n(x) - |x|\big| \le \frac{1}{\sqrt{n}} \to 0.$$

Thus f_n *converges uniformly to* $|x|$.

Each function f_n *is smooth, but the limit function* $|x|$ *is not differentiable at* $x = 0$. *This example therefore shows that uniform convergence alone is not sufficient to preserve differentiability in the limit.*

Exercise 605. *Investigate the sequence of functions*

$$f_n(x) = \sum_{k=0}^{n} \frac{x^k}{k!} \qquad (x \in [0,1])$$

with respect to uniform convergence to e^x. *Then use the theorem on uniform convergence and continuity.*

Solution 606. *For every* $x \in [0,1]$, *we have*

$$e^x = \sum_{k=0}^{\infty} \frac{x^k}{k!}.$$

Therefore

$$e^x - f_n(x) = \sum_{k=n+1}^{\infty} \frac{x^k}{k!}.$$

Since $0 \le x \le 1$, *it follows that*

$$0 \le e^x - f_n(x) \le \sum_{k=n+1}^{\infty} \frac{1}{k!}.$$

The right-hand side does not depend on x and tends to 0. Hence (f_n) converges uniformly to e^x on $[0, 1]$.

Since each partial sum f_n is a polynomial and therefore continuous, the theorem on uniform convergence and continuity implies that the limit function e^x is continuous as well.

Exercise 607. *Let*

$$f_n(x) = \frac{x}{n} \qquad (x \in [0, 1]).$$

Show that the limit and the integral may be interchanged, and compute both sides explicitly.

Solution 608. *For $x \in [0, 1]$, we have*

$$|f_n(x)| = \frac{x}{n} \leq \frac{1}{n}.$$

Hence

$$\|f_n\|_\infty \leq \frac{1}{n} \to 0.$$

Thus (f_n) converges uniformly to the zero function $f(x) = 0$.

We now compute the integrals:

$$\int_0^1 f_n(x)\,dx = \int_0^1 \frac{x}{n}\,dx = \frac{1}{n}\int_0^1 x\,dx = \frac{1}{2n}.$$

Therefore

$$\lim_{n\to\infty} \int_0^1 f_n(x)\,dx = 0.$$

On the other hand,

$$\int_0^1 f(x)\,dx = \int_0^1 0\,dx = 0.$$

Thus indeed

$$\int_0^1 f(x)\,dx = \lim_{n\to\infty} \int_0^1 f_n(x)\,dx.$$

Exercise 609. *Give an example of a sequence of continuous functions on* $[0,1]$ *that converges pointwise to the zero function, but whose integrals do not converge to the integral of the limit function. Briefly explain why this shows that pointwise convergence is not sufficient for interchanging the integral and the limit.*

Solution 610. *A classical example is given by the triangular functions*

$$f_n(x) = \begin{cases} n^2 x, & 0 \le x \le \frac{1}{n}, \\ n^2\left(\frac{2}{n} - x\right), & \frac{1}{n} \le x \le \frac{2}{n}, \\ 0, & \frac{2}{n} \le x \le 1. \end{cases}$$

Each function f_n *is continuous.*

For each fixed $x \in [0,1]$, *the following holds: for sufficiently large* n, *the point* x *lies outside the support of* f_n, *and hence* $f_n(x) = 0$. *Therefore* (f_n) *converges pointwise to the zero function.*

The area under the graph of f_n *is a triangle with base length* $\frac{2}{n}$ *and height* n. *Hence*

$$\int_0^1 f_n(x)\,dx = \frac{1}{2} \cdot \frac{2}{n} \cdot n = 1 \qquad \textit{for all } n.$$

Consequently,

$$\lim_{n\to\infty}\int_0^1 f_n(x)\,dx = 1,$$

whereas

$$\int_0^1 0\,dx = 0.$$

Thus

$$\lim_{n\to\infty}\int_0^1 f_n(x)\,dx \neq \int_0^1 \lim_{n\to\infty} f_n(x)\,dx.$$

This shows that pointwise convergence alone is not sufficient for the interchange theorem.

Exercise 611. *Let*

$$f_n(x) = \sum_{k=0}^{n} \frac{x^k}{k!} \qquad (x \in [0,1]).$$

Show that the derivatives f_n' also converge uniformly to e^x, and conclude from this that (f_n) permits differentiation in the passage to the limit.

Solution 612. *We have already seen that f_n converges uniformly to e^x. We now compute the derivatives:*

$$f_n'(x) = \sum_{k=1}^{n} \frac{kx^{k-1}}{k!} = \sum_{k=1}^{n} \frac{x^{k-1}}{(k-1)!} = \sum_{m=0}^{n-1} \frac{x^m}{m!}.$$

This is again a partial sum of the exponential series. Hence, on $[0,1]$, we likewise have

$$e^x - f_n'(x) = \sum_{k=n}^{\infty} \frac{x^k}{k!},$$

and therefore

$$0 \leq e^x - f_n'(x) \leq \sum_{k=n}^{\infty} \frac{1}{k!} \to 0.$$

Thus f_n' converges uniformly to e^x.

By the theorem on interchanging limits and derivatives, it now follows that the limit function $f(x) = e^x$ is differentiable and that

$$f'(x) = e^x.$$

Exercise 613. *Investigate the power series*

$$\sum_{n=1}^{\infty} \frac{x^n}{n^2}.$$

Determine its radius of convergence and show that the series converges absolutely and uniformly on every interval $[-r, r]$ with $0 < r < 1$. Also give the formally differentiated series.

Solution 614. *For the coefficients $a_n = \frac{1}{n^2}$, we have*

$$\sqrt[n]{|a_n|} = \sqrt[n]{\frac{1}{n^2}} \to 1.$$

Therefore the radius of convergence is

$$R = 1.$$

Now let $0 < r < 1$. For $x \in [-r, r]$, we have

$$\left|\frac{x^n}{n^2}\right| \leq \frac{r^n}{n^2} \leq r^n.$$

Since the geometric series $\sum r^n$ *converges, the Weierstrass majorant criterion implies absolute and uniform convergence on* $[-r, r]$.

The formally differentiated series is

$$\sum_{n=1}^{\infty} \frac{nx^{n-1}}{n^2} = \sum_{n=1}^{\infty} \frac{x^{n-1}}{n}.$$

This series is also convergent for $|x| < 1$, *and on every smaller closed interval one may differentiate term by term.*

Exercise 615. *Let*

$$f_n(x) = x^n \qquad (x \in [0, 1]).$$

Compute $\int_0^1 f_n(x)\,dx$ *and compare the result with the pointwise limit function. What does this example show about the endpoint* $x = 1$*?*

Solution 616. *For every* $n \in \mathbb{N}$, *we have*

$$\int_0^1 x^n\,dx = \frac{1}{n+1}.$$

Hence

$$\lim_{n\to\infty} \int_0^1 x^n\,dx = 0.$$

The pointwise limit function is

$$f(x) = \begin{cases} 0, & 0 \le x < 1, \\ 1, & x = 1. \end{cases}$$

This function is Riemann integrable on $[0,1]$, *and its integral is also* 0, *since the value at a single point does not contribute any area.*

However, the example shows that the limit value at the endpoint $x=1$ *differs from the limit values in the interior. Precisely for this reason,* (f_n) *does not converge uniformly on* $[0,1]$. *Thus the endpoint is the only place where the convergence becomes problematic.*

Chapter 18

Taylor Series

18.1 Motivation and First Orientation

In the preceding chapters, we learned to understand functions not only pointwise, but also through sequences and series. Uniform convergence was particularly important because it allows limit processes to be interchanged with continuity, integration, and differentiation. In the last section on power series, we saw that many elementary functions – such as the exponential function, sine, and cosine – can be represented by power series in the interior of their domains of convergence. Such series are more than mere computational formulas: they provide local approximations, make limits transparent, and allow precise error estimates.

The fundamental idea of this chapter is the following:

If a function is sufficiently often differentiable, then in a neighborhood of a point a it can be approximated by a polynomial whose coefficients are obtained precisely from the derivatives of f at a. The more derivatives one takes into account, the better the approximation becomes near a. This leads to the Taylor polynomials and to Taylor's formula with integral remainder.

Theorem 617 (Taylor formula). *Let $I \subset \mathbb{R}$ be a nonempty interval, let $a \in I$, and let $f : I \to \mathbb{R}$ be an $(n+1)$-times continuously differentiable function. Then, for every $x \in I$, we have*

$$\begin{aligned} f(x) = f(a) + \frac{f'(a)}{1!}(x-a) + \frac{f''(a)}{2!}(x-a)^2 + \cdots \\ + \frac{f^{(n)}(a)}{n!}(x-a)^n + R_{n+1}(x), \end{aligned}$$

where the remainder term is

$$R_{n+1}(x) = \frac{1}{n!}\int_a^x (x-t)^n f^{(n+1)}(t)\, dt$$

. *Here the integral is to be understood as an oriented integral.*

Proof. We first explain the mechanism behind the formula.

For $n = 0$, the statement is precisely the already known fundamental theorem of calculus:

$$f(x) = f(a) + \int_a^x f'(t)\, dt.$$

For the general case, for $k \geq 0$ we consider the auxiliary quantity

$$R_{k+1}(x) := \frac{1}{k!} \int_a^x (x-t)^k f^{(k+1)}(t)\, dt.$$

We prove the recursion formula

$$R_{k+1}(x) = \frac{f^{(k+1)}(a)}{(k+1)!}(x-a)^{k+1} + R_{k+2}(x).$$

To do so, we use integration by parts with

$$u(t) = f^{(k+1)}(t), \qquad dv = (x-t)^k\, dt.$$

Then

$$du = f^{(k+2)}(t)\, dt \qquad \text{and} \qquad v(t) = -\frac{(x-t)^{k+1}}{k+1}.$$

Hence we obtain

$$\begin{aligned} R_{k+1}(x) &= \frac{1}{k!} \int_a^x (x-t)^k f^{(k+1)}(t)\, dt \\ &= \frac{1}{k!} \left[-\frac{(x-t)^{k+1}}{k+1} f^{(k+1)}(t) \right]_{t=a}^{t=x} \\ &\quad + \frac{1}{(k+1)!} \int_a^x (x-t)^{k+1} f^{(k+2)}(t)\, dt \\ &= \frac{f^{(k+1)}(a)}{(k+1)!}(x-a)^{k+1} + R_{k+2}(x). \end{aligned}$$

Everything now starts with

$$f(x) = f(a) + R_1(x).$$

Using the recurrence formula, we obtain

$$\begin{aligned} f(x) &= f(a) + \frac{f'(a)}{1!}(x-a) + R_2(x) \\ &= f(a) + \frac{f'(a)}{1!}(x-a) + \frac{f''(a)}{2!}(x-a)^2 + R_3(x) \\ &\vdots \\ &= f(a) + \frac{f'(a)}{1!}(x-a) + \cdots + \frac{f^{(n)}(a)}{n!}(x-a)^n \\ &\quad + R_{n+1}(x). \end{aligned}$$

This proves Taylor's formula. □

The formula shows that the Taylor polynomial reproduces the function value and the derivatives up to order n exactly at the point a. The remainder term measures the error of the approximation. If x is close to a, this error is often small, and this is precisely where the great practical importance of the Taylor expansion lies.

18.1.1 A first consequence of Taylor's formula

Taylor's formula immediately shows that functions whose $(n+1)$-st derivative vanishes are already represented by their Taylor polynomial.

Corollary 618. *Let $I \subset \mathbb{R}$ be a nonempty interval, let $a \in I$, and let $f : I \to \mathbb{R}$ be an $(n+1)$-times continuously differentiable function. If*

$$f^{(n+1)}(x) = 0 \qquad \textit{for all } x \in I,$$

then f is a polynomial of degree at most n.

Proof. By Taylor's formula, for every $x \in I$ we have

$$f(x) = f(a) + \frac{f'(a)}{1!}(x-a) + \cdots + \frac{f^{(n)}(a)}{n!}(x-a)^n + \frac{1}{n!}\int_a^x (x-t)^n f^{(n+1)}(t)\,dt.$$

Since, by assumption, $f^{(n+1)}(t) = 0$ for all $t \in I$, the remainder term vanishes. Hence f itself is already a polynomial of degree at most n. □

18.2 The Lagrange Remainder

From the integral form of the remainder one can now obtain a particularly transparent representation of the error. It shows that the error between the function and its Taylor polynomial is described by the $(n+1)$-st derivative at a suitable point.

Theorem 619 (Lagrange remainder). *Let $I \subset \mathbb{R}$ be a nonempty interval, let $a, x \in I$, and let $f : I \to \mathbb{R}$ be an $(n+1)$-times continuously differentiable function. Then there exists a point b between a and x such that*

$$f(x) = \sum_{k=0}^{n} \frac{f^{(k)}(a)}{k!}(x-a)^k + \frac{f^{(n+1)}(b)}{(n+1)!}(x-a)^{n+1}.$$

Proof. We use Taylor's formula with integral remainder:

$$f(x) = \sum_{k=0}^{n} \frac{f^{(k)}(a)}{k!}(x-a)^k + \frac{1}{n!}\int_a^x (x-t)^n f^{(n+1)}(t)\,dt.$$

It is therefore enough to evaluate the integral in a suitable form.

For $x = a$, the statement is trivial. Thus let $x \neq a$. We first consider the case $x > a$. Then the function

$$\varphi(t) := (x - t)^n$$

is nonnegative and continuous on the interval $[a, x]$. Since, moreover, $f^{(n+1)}$ is continuous, we may apply the mean value theorem for integrals. Hence there exists a $b \in (a, x)$ such that

$$\int_a^x (x-t)^n f^{(n+1)}(t)\, dt = f^{(n+1)}(b) \int_a^x (x-t)^n\, dt.$$

Now

$$\int_a^x (x-t)^n\, dt = \left[-\frac{(x-t)^{n+1}}{n+1}\right]_{t=a}^{t=x} = \frac{(x-a)^{n+1}}{n+1}.$$

It follows that

$$\begin{aligned} \frac{1}{n!}\int_a^x (x-t)^n f^{(n+1)}(t)\, dt &= \frac{f^{(n+1)}(b)}{n!} \cdot \frac{(x-a)^{n+1}}{n+1} \\ &= \frac{f^{(n+1)}(b)}{(n+1)!}(x-a)^{n+1}. \end{aligned}$$

For $x < a$, one proceeds analogously, using the oriented integral; in this case one obtains the same expression with a point b between x and a. This proves the assertion. □

Remark 620. *The Lagrange remainder makes the approximation particularly intuitive: the error depends only on a derivative of higher order at a suitable intermediate point. If x is close to a, or if $f^{(n+1)}$ is small on the interval under consideration, then the Taylor error is small as well.*

18.3 Taylor Polynomials

The Taylor formula suggests that, for a sufficiently often differentiable function, one should consider a polynomial which, at a fixed point a, reproduces not only the function value but also the derivatives up to a certain order. Such polynomials are called *Taylor polynomials.*

Definition 621 (Taylor polynomial). *Let $I \subset \mathbb{R}$ be a nonempty interval, let $a \in I$, and let $f : I \to \mathbb{R}$ be an n-times continuously differentiable function. Then*

$$T_n(f,a)(x) := \sum_{k=0}^{n} \frac{f^{(k)}(a)}{k!}(x-a)^k$$

is called the Taylor polynomial *of f at the point a of degree at most n.*

The Taylor polynomial is characterized by the fact that it reproduces the derivatives of f at the expansion point a up to order n:

$$T_n(f,a)^{(k)}(a) = f^{(k)}(a) \qquad (k = 0, 1, \ldots, n).$$

In particular, $T_n(f, a)$ is the polynomial function that adapts f near a as closely as possible to its local derivative data.

18.4 A Consequence of the Lagrange Remainder

The Lagrange remainder yields a particularly useful form of the approximation theorem. It describes the error of the Taylor approximation as a function that vanishes relative to $(x-a)^n$.

Corollary 622. *Let $I \subset \mathbb{R}$ be a nonempty interval, let $a \in I$, and let $f : I \to \mathbb{R}$ be an n-times continuously differentiable function. Then there exists a function $g : I \to \mathbb{R}$ such that*

$$\lim_{x \to a} g(x) = 0$$

and

$$f(x) = \sum_{k=0}^{n} \frac{f^{(k)}(a)}{k!}(x-a)^k + g(x)(x-a)^n \qquad (x \in I).$$

Proof. We apply the Lagrange remainder to order $n-1$. Then, for each $x \in I$, there exists a point b between a and x such that

$$f(x) = \sum_{k=0}^{n-1} \frac{f^{(k)}(a)}{k!}(x-a)^k + \frac{f^{(n)}(b)}{n!}(x-a)^n.$$

We now add and subtract the term

$$\frac{f^{(n)}(a)}{n!}(x-a)^n.$$

It follows that

$$f(x) = \sum_{k=0}^{n} \frac{f^{(k)}(a)}{k!}(x-a)^k + \frac{f^{(n)}(b) - f^{(n)}(a)}{n!}(x-a)^n.$$

Thus, for $x \neq a$, we set

$$g(x) := \frac{f^{(n)}(b) - f^{(n)}(a)}{n!}, \qquad \text{and} \qquad g(a) := 0.$$

Since b lies between a and x, we also have $b \to a$ as $x \to a$. By the continuity of $f^{(n)}$ at a, it follows that

$$f^{(n)}(b) \to f^{(n)}(a),$$

hence

$$g(x) \to 0 \qquad (x \to a).$$

This proves the assertion. □

Remark 623. *Equivalently, this statement is often written in the form*

$$f(x) = \sum_{k=0}^{n} \frac{f^{(k)}(a)}{k!}(x-a)^k + o((x-a)^n) \qquad (x \to a).$$

This means that the error is of smaller order than $(x-a)^n$.

18.5 Examples of Taylor Expansions

We now consider some of the most important elementary functions and expand them about the point $a = 0$. This expansion point is particularly convenient because the derivatives usually take a simple form there, and the resulting series are the familiar power series of analysis.

Example 624 (The exponential function). *Let* $f(x) = e^x$. *Then, for all* $k \in \mathbb{N}_0$,

$$f^{(k)}(x) = e^x.$$

In particular,

$$f^{(k)}(0) = 1 \qquad (k \in \mathbb{N}_0).$$

Thus the Taylor polynomial of degree n *at the point* 0 *is given by*

$$T_n(e^x, 0) = \sum_{k=0}^{n} \frac{x^k}{k!}.$$

By the Lagrange form of the remainder, for every $x \in \mathbb{R}$ *there exists a point* ξ *between* 0 *and* x *such that*

$$e^x = \sum_{k=0}^{n} \frac{x^k}{k!} + \frac{e^\xi}{(n+1)!} x^{n+1}.$$

Hence the remainder satisfies

$$R_{n+1}(x) = \frac{e^\xi}{(n+1)!} x^{n+1}.$$

For fixed x, it follows that

$$|R_{n+1}(x)| \leq \frac{e^{|x|}}{(n+1)!}|x|^{n+1} \xrightarrow[n\to\infty]{} 0.$$

Thus we obtain the Taylor series

$$e^x = \sum_{k=0}^{\infty} \frac{x^k}{k!}.$$

Example 625 (The sine function). *Let $f(x) = \sin x$. The derivatives are periodic:*

$$f'(x) = \cos x, \qquad f''(x) = -\sin x,$$
$$f^{(3)}(x) = -\cos x, \qquad f^{(4)}(x) = \sin x.$$

In particular, it follows that

$$f(0) = 0, \qquad f'(0) = 1, \qquad f''(0) = 0, \qquad f^{(3)}(0) = -1,$$

and, in general,

$$f^{(2k)}(0) = 0, \qquad f^{(2k+1)}(0) = (-1)^k \qquad (k \in \mathbb{N}_0).$$

Therefore, the Taylor polynomial of $\sin x$ *at the point* 0 *is*

$$T_{2m+1}(\sin x, 0) = \sum_{k=0}^{m} (-1)^k \frac{x^{2k+1}}{(2k+1)!}.$$

By the Lagrange form of the remainder, for every $x \in \mathbb{R}$ there is a point ξ between 0 *and x such that*

$$\sin x = \sum_{k=0}^{m} (-1)^k \frac{x^{2k+1}}{(2k+1)!} + \frac{f^{(2m+2)}(\xi)}{(2m+2)!} x^{2m+2}.$$

Since every derivative of sin *has absolute value at most* 1, *we have*

$$|R_{2m+2}(x)| \leq \frac{|x|^{2m+2}}{(2m+2)!}.$$

For fixed x*, this expression tends to* 0*, and hence the Taylor series is*

$$\sin x = \sum_{k=0}^{\infty} (-1)^k \frac{x^{2k+1}}{(2k+1)!}.$$

Example 626 (The cosine function). *For* $f(x) = \cos x$*, the derivatives likewise follow a periodic pattern:*

$$\begin{aligned} f'(x) &= -\sin x, & f''(x) &= -\cos x, \\ f^{(3)}(x) &= \sin x, & f^{(4)}(x) &= \cos x. \end{aligned}$$

In particular,

$$f(0) = 1, \qquad f'(0) = 0, \qquad f''(0) = -1, \qquad f^{(3)}(0) = 0,$$

and, in general,

$$f^{(2k)}(0) = (-1)^k, \qquad f^{(2k+1)}(0) = 0 \qquad (k \in \mathbb{N}_0).$$

Hence the Taylor polynomial of degree $2m$ *about* 0 *is*

$$T_{2m}(\cos x, 0) = \sum_{k=0}^{m} (-1)^k \frac{x^{2k}}{(2k)!}.$$

Again, the Lagrange form of the remainder gives a point ξ *between* 0 *and* x *such that*

$$\cos x = \sum_{k=0}^{m} (-1)^k \frac{x^{2k}}{(2k)!} + \frac{f^{(2m+1)}(\xi)}{(2m+1)!} x^{2m+1}.$$

Since here as well every derivative has absolute value at most 1*, it follows that*

$$|R_{2m+1}(x)| \leq \frac{|x|^{2m+1}}{(2m+1)!} \xrightarrow[m\to\infty]{} 0.$$

Thus we obtain the Taylor series

$$\cos x = \sum_{k=0}^{\infty} (-1)^k \frac{x^{2k}}{(2k)!}.$$

Example 627 (The geometric series as a Taylor series)**.** *Let* $f(x) = \frac{1}{1-x}$ *for* $x < 1$*. Then, for all* $k \in \mathbb{N}_0$*,*

$$f^{(k)}(x) = \frac{k!}{(1-x)^{k+1}}.$$

In particular,

$$f^{(k)}(0) = k!.$$

Thus the Taylor polynomial about 0 *is given by*

$$T_n\left(\frac{1}{1-x}, 0\right) = \sum_{k=0}^{n} x^k.$$

Indeed, for every $n \in \mathbb{N}_0$*, the identity*

$$\frac{1}{1-x} = \sum_{k=0}^{n} x^k + \frac{x^{n+1}}{1-x}, \qquad x \neq 1$$

holds. For $|x| < 1$*, it follows that*

$$\frac{x^{n+1}}{1-x} \xrightarrow[n\to\infty]{} 0,$$

and hence

$$\frac{1}{1-x} = \sum_{k=0}^{\infty} x^k \qquad (|x| < 1).$$

Example 628 (The logarithm). *For $f(x) = \ln(1+x)$ we consider the expansion point $a = 0$. We have*

$$f'(x) = \frac{1}{1+x}, \qquad f''(x) = -\frac{1}{(1+x)^2},$$
$$f^{(3)}(x) = \frac{2!}{(1+x)^3}.$$

and, in general,

$$f^{(k)}(x) = (-1)^{k-1}\frac{(k-1)!}{(1+x)^k} \qquad (k \geq 1).$$

Therefore,

$$f(0) = 0, \qquad f^{(k)}(0) = (-1)^{k-1}(k-1)! \qquad (k \geq 1).$$

Thus the Taylor polynomial of degree n at the point 0 is

$$T_n(\ln(1+x), 0) = \sum_{k=1}^{n} (-1)^{k-1}\frac{x^k}{k}.$$

For $|x| < 1$, this gives the well-known Taylor series

$$\ln(1+x) = \sum_{k=1}^{\infty} (-1)^{k-1}\frac{x^k}{k}.$$

These examples show how the abstract Taylor formula turns into very concrete and computable expansion formulas. At the same time, one sees that the Taylor series of the elementary functions do not arise by chance, but follow directly from the derivative values at the expansion point.

18.6 Power Series as Taylor Series

In Chapter 16 we saw that power series may be differentiated and integrated term by term in the interior of their interval of convergence. This naturally raises the question of how power series are related to Taylor series. The answer is particularly elegant: in the interior of its interval of convergence, a power series is already its own Taylor series.

Theorem 629. *Let*

$$f(x) = \sum_{n=0}^{\infty} b_n (x-a)^n$$

be a power series with positive radius of convergence r. Then, for every $x \in]a-r, a+r[$, we have

$$f^{(n)}(a) = n!\, b_n \qquad (n \in \mathbb{N}_0).$$

In particular, the Taylor series of f about the expansion point a is exactly the given power series, that is,

$$f(x) = \sum_{n=0}^{\infty} \frac{f^{(n)}(a)}{n!}(x-a)^n \qquad (|x-a| < r).$$

Proof. By the theorem on termwise differentiation of power series, f is infinitely often differentiable on the open interval $]a-r, a+r[$, and for every $n \in \mathbb{N}_0$ the power series may be differentiated term by term n times. This gives

$$f^{(n)}(x) = \sum_{k=n}^{\infty} b_k\, k(k-1)\cdots(k-n+1)(x-a)^{k-n}.$$

Now setting $x = a$, all summands with $k > n$ vanish, and only the term with $k = n$ remains:

$$f^{(n)}(a) = b_n\, n!.$$

Thus the Taylor coefficients of f at the point a are precisely determined by the coefficients of the power series. Hence the Taylor series of f is exactly the original power series. □

Remark 630. *In particular, this theorem explains why the classical series for* e^x, $\sin x$, $\cos x$ *and* $\ln(1+x)$ *are not merely accidental expansions: they are exactly the Taylor series of the corresponding functions.*

18.7 Abel's Limit Theorem

In the chapter on power series, we have already seen that power series are very well behaved inside their radius of convergence. The following theorem now shows that a power series also behaves particularly well at the boundary point $x = 1$, provided that its series of coefficients converges.

Theorem 631 (Abel's limit theorem). *Let*

$$\sum_{n=0}^{\infty} b_n$$

be a convergent series of real numbers, and let

$$f(x) := \sum_{n=0}^{\infty} b_n x^n \qquad (x \in [0,1]).$$

Then the power series converges uniformly on $[0,1]$. *In particular,* f *is continuous on* $[0,1]$, *and*

$$f(1) = \sum_{n=0}^{\infty} b_n \qquad \text{and therefore} \qquad \lim_{x \to 1^-} f(x) = \sum_{n=0}^{\infty} b_n.$$

Proof. Let

$$s_n := \sum_{k=0}^{n} b_k$$

be the sequence of partial sums of the convergent series $\sum b_n$, and let

$$s := \sum_{n=0}^{\infty} b_n$$

be its limit. Then

$$c_n := s_n - s \longrightarrow 0.$$

We now show that the partial sums of the power series are uniformly Cauchy on $[0, 1]$.

For $M > N \geq 1$ and $x \in [0, 1]$, consider the remainder

$$\sum_{n=N}^{M} b_n x^n.$$

Since, for $n \geq 1$,

$$b_n = s_n - s_{n-1} = c_n - c_{n-1},$$

partial summation gives

$$\begin{aligned}\sum_{n=N}^{M} b_n x^n &= \sum_{n=N}^{M} (c_n - c_{n-1}) x^n \\ &= c_M x^M - c_{N-1} x^N + (1 - x) \sum_{n=N}^{M-1} c_n x^n.\end{aligned}$$

Hence we obtain the estimate

$$\left| \sum_{n=N}^{M} b_n x^n \right| \leq |c_M| + |c_{N-1}| + (1 - x) \sum_{n=N}^{M-1} |c_n| x^n.$$

Now let $\varepsilon > 0$. Since $c_n \to 0$, there exists N_0 such that

$$|c_n| < \varepsilon \qquad \text{for all } n \geq N_0 - 1.$$

Thus, for $M > N \geq N_0$ and all $x \in [0, 1]$, we get

$$\left| \sum_{n=N}^{M} b_n x^n \right| \leq \varepsilon + \varepsilon + \varepsilon (1 - x) \sum_{n=N}^{M-1} x^n.$$

Since

$$(1-x)\sum_{n=N}^{M-1} x^n \leq 1,$$

we finally obtain

$$\left|\sum_{n=N}^{M} b_n x^n\right| \leq 3\varepsilon \qquad \text{for all } x \in [0,1].$$

Thus the series is uniformly Cauchy on $[0,1]$ and therefore converges uniformly.

Each partial sum

$$p_M(x) := \sum_{n=0}^{M} b_n x^n$$

is a polynomial and hence continuous. Therefore the uniform limit f is also continuous on $[0,1]$.

Finally, for $x=1$ we have

$$f(1) = \sum_{n=0}^{\infty} b_n,$$

and so, by the continuity of f on $[0,1]$,

$$\lim_{x\to 1^-} f(x) = f(1) = \sum_{n=0}^{\infty} b_n.$$

□

Remark 632. *The theorem gives an intuitive explanation of why power series often remain well controlled near a boundary point. For Taylor series this is especially important, because it allows one to describe limiting processes and function values at boundary points in a precise way.*

18.8 Exercises with Solutions

Exercise 633. *Determine the Taylor polynomial of $f(x) = e^x$ at the point $a = 0$ up to order 3. Also give the remainder term in Lagrange form and estimate $|R_4(0.1)|$.*

Solution 634. *For $f(x) = e^x$, we have for all $k \in \mathbb{N}_0$:*

$$f^{(k)}(x) = e^x.$$

In particular, $f^{(k)}(0) = 1$. Hence the Taylor polynomial of order 3 is

$$T_3(e^x, 0) = 1 + x + \frac{x^2}{2!} + \frac{x^3}{3!}.$$

By the Lagrange form of the remainder, for each x there is a point ξ between 0 and x such that

$$R_4(x) = \frac{e^\xi}{4!}x^4.$$

For $x = 0.1$, since $0 \le \xi \le 0.1$, it follows that

$$|R_4(0.1)| \le \frac{e^{0.1}}{24} \cdot 0.1^4.$$

This gives the desired estimate.

Exercise 635. *Determine the Taylor polynomial of* $f(x) = \sin x$ *at the point* $a = 0$ *up to order* 5.

Solution 636. *The derivatives of* $\sin x$ *are periodic:*

$$\sin x, \ \cos x, \ -\sin x, \ -\cos x, \ \sin x, \ldots$$

Therefore,

$$\begin{array}{ccc} f(0) = 0, & f'(0) = 1, & f''(0) = 0, \\ f^{(3)}(0) = -1, & f^{(4)}(0) = 0, & f^{(5)}(0) = 1. \end{array}$$

Thus,

$$T_5(\sin x, 0) = x - \frac{x^3}{3!} + \frac{x^5}{5!}.$$

Exercise 637. *Compute the limit*

$$\lim_{x \to 0} \frac{e^x - 1 - x}{x^2}.$$

Solution 638. *From the Taylor expansion of* e^x *about* 0, *it follows that*

$$e^x = 1 + x + \frac{x^2}{2} + R_3(x), \qquad \text{where} \qquad \frac{R_3(x)}{x^2} \to 0 \quad (x \to 0).$$

Thus

$$\frac{e^x - 1 - x}{x^2} = \frac{1}{2} + \frac{R_3(x)}{x^2}.$$

Hence

$$\lim_{x \to 0} \frac{e^x - 1 - x}{x^2} = \frac{1}{2}.$$

Exercise 639. *Show, using Taylor's formula, that*

$$\lim_{x\to 0} \frac{\sin x - x}{x^3} = -\frac{1}{6}.$$

Solution 640. *For* $\sin x$*, the Taylor expansion about* 0 *is*

$$\sin x = x - \frac{x^3}{3!} + R_5(x).$$

Therefore,

$$\sin x - x = -\frac{x^3}{6} + R_5(x).$$

Dividing by x^3 *yields*

$$\frac{\sin x - x}{x^3} = -\frac{1}{6} + \frac{R_5(x)}{x^3}.$$

By the estimate for the Taylor remainder, $\frac{R_5(x)}{x^3} \to 0$ *as* $x \to 0$*. Hence*

$$\lim_{x\to 0} \frac{\sin x - x}{x^3} = -\frac{1}{6}.$$

Exercise 641. *Determine the Taylor expansion of*

$$f(x) = \ln(1+x)$$

about the point 0 *up to order* 3.

Solution 642. *We have*

$$f'(x) = \frac{1}{1+x}, \qquad f''(x) = -\frac{1}{(1+x)^2},$$
$$f^{(3)}(x) = \frac{2}{(1+x)^3}.$$

Thus, at $x = 0$, we obtain

$$f(0) = 0, \qquad f'(0) = 1, \qquad f''(0) = -1, \qquad f^{(3)}(0) = 2.$$

Hence the Taylor polynomial of order 3 is

$$T_3(\ln(1+x), 0) = x - \frac{x^2}{2} + \frac{x^3}{3}.$$

Exercise 643. *Let*

$$f(x) = \sum_{n=0}^{\infty} b_n (x-a)^n$$

be a power series with positive radius of convergence. Show that

$$f^{(n)}(a) = n!\, b_n \qquad (n \in \mathbb{N}_0).$$

Solution 644. *Since the power series is termwise differentiable in the interior of its radius of convergence, we may differentiate term by term n times. It follows that*

$$f^{(n)}(x) = \sum_{k=n}^{\infty} b_k\, k(k-1)\cdots(k-n+1)(x-a)^{k-n}.$$

Setting $x = a$, all terms with $k > n$ vanish, since they contain a factor $(x-a)$ with positive exponent. Only the term $k = n$ remains:

$$f^{(n)}(a) = b_n\, n!.$$

This proves the assertion.

Exercise 645. *Show that, for all $x \in \mathbb{R}$,*

$$\left| e^x - \left(1 + x + \frac{x^2}{2}\right) \right| \le \frac{e^{|x|}}{6}|x|^3.$$

Solution 646. *We use Taylor's formula for $f(x) = e^x$ about 0 up to second order:*

$$e^x = 1 + x + \frac{x^2}{2} + \frac{e^\xi}{6}x^3$$

for some ξ between 0 and x. Thus the remainder is

$$R_3(x) = \frac{e^\xi}{6}x^3.$$

Since ξ lies between 0 and x, we have $|\xi| \le |x|$, hence

$$e^\xi \le e^{|\xi|} \le e^{|x|}.$$

Therefore,

$$\left| e^x - \left(1 + x + \frac{x^2}{2}\right) \right| = |R_3(x)| \le \frac{e^{|x|}}{6}|x|^3.$$

Exercise 647. *Compute the limit*

$$\lim_{x \to 0} \frac{1 - \cos x}{x^2}.$$

Solution 648. *We expand $\cos x$ about 0:*

$$\cos x = 1 - \frac{x^2}{2} + \frac{f^{(4)}(\xi)}{4!}x^4$$

for some ξ between 0 and x. Hence

$$1 - \cos x = \frac{x^2}{2} - R_4(x),$$

where $R_4(x) = O(x^4)$ as $x \to 0$. Dividing by x^2 gives

$$\frac{1 - \cos x}{x^2} = \frac{1}{2} - \frac{R_4(x)}{x^2}.$$

Since $\frac{R_4(x)}{x^2} \to 0$, it follows that

$$\lim_{x \to 0} \frac{1 - \cos x}{x^2} = \frac{1}{2}.$$

Exercise 649. *Show that for all $x \in \mathbb{R}$,*

$$|\sin x - x| \leq \frac{|x|^3}{6}.$$

Solution 650. *The Taylor expansion of $\sin x$ about 0 up to first order is*

$$\sin x = x + \frac{f^{(3)}(\xi)}{3!} x^3$$

for some ξ between 0 and x. Here $f^{(3)}(t) = -\cos t$, hence

$$|f^{(3)}(\xi)| = |\cos \xi| \leq 1.$$

Therefore

$$|\sin x - x| = \left| \frac{f^{(3)}(\xi)}{6} x^3 \right| \leq \frac{|x|^3}{6}.$$

This proves the estimate.

Exercise 651. *Compute the limit*

$$\lim_{x\to 0} \frac{\ln(1+x) - x + \frac{x^2}{2}}{x^3}.$$

Solution 652. *We use the Taylor expansion of* $\ln(1+x)$ *about* 0 *up to second order:*

$$\ln(1+x) = x - \frac{x^2}{2} + \frac{f^{(3)}(\xi)}{3!}x^3$$

for some ξ *between* 0 *and* x. *Since* $f^{(3)}(t) = \frac{2}{(1+t)^3}$, *it follows that*

$$\ln(1+x) - x + \frac{x^2}{2} = \frac{f^{(3)}(\xi)}{6}x^3.$$

Thus

$$\frac{\ln(1+x) - x + \frac{x^2}{2}}{x^3} = \frac{f^{(3)}(\xi)}{6}.$$

Because $\xi \to 0$ *as* $x \to 0$ *and* $f^{(3)}$ *is continuous, we have*

$$f^{(3)}(\xi) \to f^{(3)}(0) = 2.$$

Consequently,

$$\lim_{x\to 0} \frac{\ln(1+x) - x + \frac{x^2}{2}}{x^3} = \frac{2}{6} = \frac{1}{3}.$$

Exercise 653. *Show that for all* x *with* $|x| \leq 1$ *one has:*

$$\left| e^x - \sum_{k=0}^{n} \frac{x^k}{k!} \right| \leq \frac{e}{(n+1)!}|x|^{n+1}.$$

Solution 654. *By Taylor's formula with the Lagrange remainder, we have*

$$e^x = \sum_{k=0}^{n} \frac{x^k}{k!} + \frac{e^\xi}{(n+1)!} x^{n+1}$$

for some ξ *between* 0 *and* x. *Therefore,*

$$\left| e^x - \sum_{k=0}^{n} \frac{x^k}{k!} \right| = \frac{e^\xi}{(n+1)!} |x|^{n+1}.$$

Since $|x| \leq 1$ *and* ξ *lies between* 0 *and* x, *it follows that* $|\xi| \leq 1$, *hence*

$$e^\xi \leq e^{|\xi|} \leq e.$$

Thus we obtain

$$\left| e^x - \sum_{k=0}^{n} \frac{x^k}{k!} \right| \leq \frac{e}{(n+1)!} |x|^{n+1}.$$

Exercise 655. *Use a Taylor expansion to show that*

$$\lim_{x \to 0} \frac{\sqrt{1+x} - 1 - \frac{x}{2}}{x^2} = -\frac{1}{8}.$$

Solution 656. *We consider* $f(x) = \sqrt{1+x}$ *and expand around* 0 *up to second order:*

$$f(x) = f(0) + f'(0)x + \frac{f''(\xi)}{2} x^2$$

for some ξ *between* 0 *and* x. *We have*

$$f(0) = 1, \qquad f'(0) = \frac{1}{2}, \qquad f''(t) = -\frac{1}{4}(1+t)^{-3/2}.$$

It follows that

$$\sqrt{1+x}-1-\frac{x}{2}=\frac{f''(\xi)}{2}x^2.$$

Dividing by x^2 gives

$$\frac{\sqrt{1+x}-1-\frac{x}{2}}{x^2}=\frac{f''(\xi)}{2}.$$

Since $\xi \to 0$ as $x \to 0$ and f'' is continuous, we have

$$f''(\xi) \to f''(0) = -\frac{1}{4}.$$

Hence

$$\lim_{x\to 0}\frac{\sqrt{1+x}-1-\frac{x}{2}}{x^2}=\frac{1}{2}\cdot\left(-\frac{1}{4}\right)=-\frac{1}{8}.$$

Exercise 657. *Show that*

$$\sin x = x + o(x) \qquad (x \to 0).$$

Solution 658. *We use the consequence of Lagrange's remainder term for the function*

$$f(x)=\sin x$$

at the expansion point $a=0$ and with $n=1$.

Since

$$f(0)=0 \qquad \textit{and} \qquad f'(0)=1,$$

the first-order Taylor polynomial is

$$T_1(x) = x.$$

By the proved consequence of Taylor's formula, there exists a function g such that

$$\lim_{x \to 0} g(x) = 0$$

and

$$\sin x = T_1(x) + g(x)(x - 0)$$

holds.

Substituting $T_1(x) = x$ gives

$$\sin x = x + g(x)x.$$

Since $g(x) \to 0$ as $x \to 0$, it follows from the definition of Landau notation that

$$\sin x = x + o(x).$$

Alternatively, the same statement follows directly from the Taylor expansion

$$\sin x = x - \frac{x^3}{6} + o(x^3).$$

Indeed, then

$$\sin x - x = -\frac{x^3}{6} + o(x^3).$$

Division by x yields

$$\frac{\sin x - x}{x} = -\frac{x^2}{6} + o(x^2) \longrightarrow 0,$$

and hence again

$$\sin x = x + o(x).$$

Exercise 659. *Consider the series*

$$\sum_{n=1}^{\infty} \frac{(-1)^{n+1}}{n}.$$

Let the associated power series be

$$f(x) = \sum_{n=1}^{\infty} \frac{(-1)^{n+1}}{n} x^n.$$

Show that f converges uniformly on the interval $[0, 1]$, and determine

$$\lim_{x \to 1^-} f(x).$$

Solution 660. *We first consider the coefficient series*

$$\sum_{n=1}^{\infty} \frac{(-1)^{n+1}}{n}.$$

This is the alternating harmonic series.

The sequence

$$a_n = \frac{1}{n}$$

is monotonically decreasing and satisfies

$$a_n \to 0.$$

Hence, by the Leibniz criterion, the series converges.

Thus the assumptions of Abel's limit theorem are satisfied. Consequently, the power series

$$f(x) = \sum_{n=1}^{\infty} \frac{(-1)^{n+1}}{n} x^n$$

converges uniformly on the interval $[0, 1]$.

In particular, f *is continuous on* $[0, 1]$, *and we have*

$$f(1) = \sum_{n=1}^{\infty} \frac{(-1)^{n+1}}{n}.$$

From Chapter 7 it is already known that the alternating harmonic series has the value

$$\ln 2.$$

Therefore it follows that

$$f(1) = \ln 2.$$

Because f *is continuous on* $[0, 1]$, *we finally obtain*

$$\lim_{x \to 1^-} f(x) = f(1) = \ln 2.$$

Abel's limit theorem therefore makes it possible to determine the limit of the power series at the endpoint $x = 1$ *from the value of the underlying coefficient series.*

Chapter 19

Concluding Remarks

19.1 Summary

This book has traced a path through the fundamental concepts and methods of analysis, leading from the first logical and algebraic prerequisites to the classical central topics of the subject. The guiding idea of the presentation was not merely to place individual theorems one after another, but to make visible the inner order and interconnection of the concepts.

The starting point was mathematical induction together with some basic combinatorial tools. These belong to the simplest, but at the same time to the most important, methods of proof in mathematics. Building on this foundation, the field axioms and the order axioms of the real numbers were developed. Through them, the structural foundation on which all further analysis rests came into view. The absolute value, the Archimedean axiom, and the resulting inequalities and order relations deepened this approach.

A subsequent focus was the theory of sequences and limits. Here it became clear how precisely the concept of convergence must be formulated and how closely it is connected with the properties of the real numbers. Limit laws, divergence to $+\infty$ and $-\infty$, and the relation between convergence and boundedness formed essential components of this discussion. With Cauchy sequences, the nested interval principle, and subsequences, the completeness axiom finally moved to the foreground: precisely that property of the real numbers which distinguishes them from the rational numbers.

The following chapters on supremum and infimum, as well as on the existence of square roots, showed how concrete mathematical results arise from abstract completeness properties. Heron's method also made clear that, in analysis, theory and procedure are closely connected.

In the chapter on series, the most important convergence criteria were developed, in particular the Cauchy criterion, the geometric series, alternating series, absolute convergence, and rearrangements. This opened up another central area of analysis, in which infinite sums are studied not merely formally, but in a precise mathematical sense.

The discussion of point sets prepared the transition to functions and continuity. First, functions were introduced as mathematical objects; then function limits, continuity, the intermediate value theorem, extrema, and uniform continuity were treated. These results form the basis for numerous later applications and show how strongly analytic statements depend on the interplay between limit concepts and order properties.

Another focus was placed on the elementary functions. Power, root, exponential, logarithmic, and trigonometric functions were studied not merely as objects of computation, but as functions connected by a coherent structure. Landau notation supplemented this section with an important tool for describing asymptotic behavior.

Differentiation made it possible to describe the local variability of functions with precision. The derivative, introduced as the limit of the difference quotient, led to a fundamental concept that reappears throughout the subsequent chapters. Local extrema, the mean value theorem, monotonicity, and l'Hospital's rules showed how powerful the derivative is as an analytic tool.

Integration introduced another central concept of analysis. Starting from step functions and upper and lower integrals, Riemann integrability was developed. The fundamental theorem of calculus then connected the two major areas of analysis in a particularly elegant way. Substitution, integration by parts, partial fraction decomposition, and trigonometric integrals expanded the repertoire of integration methods, while improper integrals directed attention to unbounded intervals and singular integrands.

The treatment of uniform convergence made clear that, for sequences of functions, not only the pointwise limit but also the mode of convergence is essential. Uniform convergence is the appropriate condition for controlling continuity, integration, and differentiation under passage to the limit. This idea became especially transparent in the study of power series: they provide a central class of functions whose convergence behavior can be described by a radius of convergence and whose term-by-term differentiation and integration are possible under suitable conditions.

The discussion concluded with Taylor series. They combine local approximation, higher-order derivatives, and series expansions into one of the most important tools of analysis. Taylor's formula with the Lagrange form of the remainder shows how functions can be described precisely by polynomials near a point of expansion. Abel's limit theorem rounds off this discussion and points to the deep connection between series, limits, and the development of functions.

This completes a coherent arc: from elementary proof techniques, through the structure of the real numbers and the theory of limits, to continuity, differentiation, integration, and series expansions. In this development, analysis does not appear as a loose collection of individual theorems, but as an internally ordered theory in which the concepts build on one another and complement one another.

Readers who master the material presented here have

a solid foundation for further study in mathematics and its applications in neighboring disciplines. At the same time, the conceptual and methodological understanding developed in this book is a prerequisite for many advanced fields in which analytic ways of thinking play a central role.

This book brings the first systematic journey through analysis to an end, but not the engagement with analysis itself. Much of what has been prepared here will be taken up again, generalized, and deepened in later courses and in more advanced literature.

19.2 Outlook

The fundamental concepts developed in this book by no means exhaust the study of analysis. Rather, they form the starting point for higher analysis, especially for the differential and integral calculus of several variables. In that setting, several coordinates take the place of a single variable, and many of the concepts and theorems introduced here reappear in natural generalizations.

The transition to this broader point of view requires a certain additional degree of abstraction. The topology of metric spaces plays a central role in this development. A metric space is, first of all, a set on which a distance between any two points is defined. From this distance one can derive open balls, convergence of sequences, continuity, and other basic notions. In this way, the language of analysis is transferred from the special structure of the real numbers to more general spaces.

This is precisely where the internal coherence of analysis becomes visible. Many of the theorems proved in this book return in generalized form in metric spaces. Concepts such as limits, continuity, completeness, compactness, and uniform continuity then no longer appear only in the context of real numbers or real-valued functions, but as structural notions within a broader theory. What was first developed here for sequences, functions, and integrals gains greater scope and new possibilities of application in that setting.

This perspective is especially important for linear algebra, numerical analysis, and probability theory. In linear algebra, metric and normed structures occur naturally, for instance in vector spaces, linear mappings, and norms. Numerical analysis uses analytic convergence and approximation properties to develop and assess computational methods. In probability theory, notions of convergence, measure spaces, and integrals play a fundamental role, so that many of the ideas introduced here reappear within an extended framework.

These ideas also lead further to functional analysis, differential geometry, and differential equations. Functional analysis studies infinite-dimensional spaces and the linear operators acting on them. Differential geometry connects analysis and geometry by describing smooth structures on manifolds. Differential equations, finally, model processes of change in the natural sciences, engineering, and many other fields of application, and they form one of the most important bridges between theory and practice.

Measure theory should also be mentioned, since it provides a more precise and more general foundation for integration. It opens the way to Lebesgue integration and is an essential prerequisite for large parts of modern analysis. Many of the integration concepts and limiting processes treated in this book appear in that framework in a refined and at the same time very powerful form.

Thus it becomes clear that analysis, as developed in this book, is not merely a completed first stage of study, but also a foundation for numerous advanced fields. Those who have understood its basic principles will recognize many familiar ideas in later courses—now in greater generality, with a new language and a broader point of view.

19.3 Book Recommendations

Analysis does not stand in isolation within mathematics, but is closely connected with several foundational disciplines. This becomes particularly clear in formal logic, linear algebra, and computer science. The following books may therefore serve as useful complements to the present work and, together with it, form a coordinated learning series for students in their first semesters.

Formal Logic – Propositional Logic, Predicate Logic, Modal Logic, and Non-Classical Logics

by Lucien Sina

Formal logic forms an essential foundation of mathematical thinking. Concepts such as statement, proof, rule of inference, equivalence, quantifier, and formal structure also play a central role in analysis. Many of the forms of argument used in this book – such as complete induction, contraposition, proofs by contradiction, and equivalence transformations – rest on basic logical principles, a secure command of which considerably facilitates the understanding of mathematical texts.

The connection between logic and analysis is therefore especially fruitful: while analysis develops precise notions of limits, continuity, and convergence, formal logic provides the linguistic and structural tools needed to formulate and prove mathematical statements rigorously. Those who study both areas together gain not only technical confidence, but also a deeper understanding of the structure of mathematical arguments.

Moreover, formal logic is an important foundation for theoretical computer science, automata theory, formal verification, and other areas in which precise reasoning and exact formalization are indispensable.

Algorithms and Data Structures

by Lucien Sina

This book offers an introduction to central concepts

of computer science, such as algorithms, data structures, runtime analysis, and efficient problem-solving methods. This field is also of great importance for mathematically oriented courses of study, since many analytical and numerical methods must be understood and implemented algorithmically.

Analysis provides essential foundations for this. Limits, series, approximations, numerical methods, and questions of convergence play a central role in algorithmic mathematics. Conversely, thinking in clear structures and efficient procedures also helps to grasp analytical content systematically. The connection between the two books is therefore particularly useful for students of computer science, mathematics, physics, and engineering.

At the same time, many applications show that analytical methods and algorithmic thinking complement one another: for example in numerical mathematics, computer graphics, optimization, cryptography, artificial intelligence, and machine learning.

Linear Algebra and Analytic Geometry

by Lucien Sina

This textbook offers a systematic introduction to linear algebra and analytic geometry at university level. Step by step, it develops central topics such as vector spaces, linear mappings, matrices, determinants, eigenvalue theory, as well as Euclidean and unitary vector spaces.

For analysis, this field is an ideal complement in

several respects. On the one hand, linear algebra broadens the understanding of structures, mappings, and the treatment of abstract spaces. On the other hand, it provides an essential foundation for multivariable analysis, the later differential and integral calculus of several variables, as well as many applications in physics, engineering, and numerical mathematics.

At the same time, linear algebra is already used in a fundamental way in this book on analysis. Even the treatment of sequences, series, function spaces, questions of convergence, and linear structures shows that linear thinking does not become important only in later courses, but already plays a supporting role here. Those who master linear algebra alongside analysis understand many statements not only more quickly, but also more deeply.

The combination of the two books therefore conveys not only two central foundational disciplines, but also a coherent understanding of mathematical methods: from the first logical proofs and the structure of the real numbers to linear spaces, functions, and analytic procedures.

19.4 The Learning Series

The books mentioned are designed to complement one another and to form a self-contained learning series. Together, they gradually introduce the mathematical way of thinking and at the same time cover several of the most important foundational subjects for mathematically and computationally oriented degree programs.

The starting point is *Formal Logic.* It trains the ability to formulate statements precisely, to follow proofs, and to structure mathematical arguments cleanly. Building on this, *Analysis* introduces the theory of real numbers, sequences, series, functions, continuity, differentiation, and integration. It shows how a coherent mathematical theory develops from precise fundamental concepts.

Linear Algebra and Analytic Geometry extends this perspective to linear structures, vector spaces, and geometric relationships. In doing so, it not only provides important mathematical tools, but also lays the foundations for many more advanced areas. In addition, *Algorithms and Data Structures* presents the computer-science side of structured thinking and connects mathematical clarity with practical problem solving.

It is precisely in combination that these books reveal their particular strength: logic promotes proof-writing, analysis develops the understanding of limiting processes and continuity, linear algebra fosters thinking in structures and spaces, and computer science supports the algorithmic implementation of problems. Together, they form a solid foundation for further study and open the door to numerous later topics such as numerical analysis, differential equations, probability theory, functional analysis, measure theory, and differential geometry.

Anyone who works through this series systematically gains not only a broad foundation in content, but also a coherent view of mathematics and its applications.

Acknowledgements

My thanks go to all readers who engage with the contents of this book and accompany the journey through analysis with interest, perseverance, and curiosity. Especially in a subject such as analysis, patience, careful reading, and the willingness to think things through repeatedly are of particular importance.

If this book helps readers see mathematical connections more clearly, understand them with greater confidence, and gain easier access to more advanced topics, then it has fulfilled its purpose.

Special thanks are due to all those who contribute to the improvement of future editions through questions, comments, suggestions, or constructive criticism. Despite great care, errors, inaccuracies, or opportunities for improvement can never be ruled out entirely. Feedback and suggestions for improvement are therefore always warmly welcome.

Bibliography

[1] Stephen Abbott: *Understanding Analysis.* Springer.

[2] Tom M. Apostol: *Mathematical Analysis.* Addison-Wesley.

[3] Gerald B. Folland: *Real Analysis: Modern Techniques and Their Applications.* Wiley.

[4] Otto Forster: *Analysis 1.* Springer.

[5] Otto Forster: *Analysis 2.* Springer.

[6] Harro Heuser: *Lehrbuch der Analysis. Teil 1.* Teubner / Vieweg.

[7] Harro Heuser: *Lehrbuch der Analysis. Teil 2.* Teubner / Vieweg.

[8] Konrad Königsberger: *Analysis 1.* Springer.

[9] Konrad Königsberger: *Analysis 2.* Springer.

[10] Serge Lang: *Real and Functional Analysis.* Springer.

[11] H. L. Royden and Patrick M. Fitzpatrick: *Real Analysis*. Pearson.

[12] Walter Rudin: *Principles of Mathematical Analysis*. McGraw-Hill.

[13] Terence Tao: *Analysis I*. Springer.

[14] Terence Tao: *Analysis II*. Springer.

[15] Wolfgang Walter: *Analysis 1*. Springer.

[16] Wolfgang Walter: *Analysis 2*. Springer.

Copyright Page

Bibliographic information from the German National Library:
The German National Library lists this publication in the German National Bibliography; detailed bibliographic data can be accessed online at `dnb.dnb.de`.

Publishing:

BoD · Books on Demand GmbH
Überseering 33
22297 Hamburg
bod@bod.de

Printing:
Libri Plureos GmbH
Friedensallee 273
22763 Hamburg

ISBN: 978-3-6963-8828-7